22⁹⁵

AUSTRALIAN MATHEMATICAL SOCIETY LECTURE SERIES

Editor-in-Chief: Professor S.A. Morris, Department of Mathematics, Statistics and Computing Science, The University of New England, Armidale N.S.W. 2351, Australia

Subject Editors:
Professor C.J. Thompson, Department of Mathematics, University of Melbourne, Parkville, Victoria 3052, Australia
Professor C.C. Heyde, Department of Statistics, University of Melbourne, Parkville, Victoria 3052, Australia
Professor J.H. Loxton, Department of Pure Mathematics, University of New South Wales, Kensington, New South Wales 2033, Australia

Australian Mathematical Society Lecture Series. 6

The Mathematics
of Projectiles
in Sport

Neville de Mestre

School of Information and Computing Sciences,
Bond University, Queensland, Australia

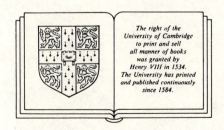

The right of the
University of Cambridge
to print and sell
all manner of books
was granted by
Henry VIII in 1534.
The University has printed
and published continuously
since 1584.

CAMBRIDGE UNIVERSITY PRESS
Cambridge
New York   Port Chester
Melbourne   Sydney

Published by the Press Syndicate of the University of Cambridge
The Pitt Building, Trumpington Street, Cambridge CB2 1RP
40 West 20th Street, New York, NY 10011, USA
10 Stamford Road, Oakleigh, Melbourne 3166, Australia

First published 1990

Printed in Great Britain at the University Press, Cambridge

British Library cataloguing in publication data available

Library of Congress cataloguing in publication data available

ISBN 0 521 39857 6 paperback

CONTENTS

Preface

A large number of sporting events contain the motion of a projectile; yet research papers have appeared only intermittently on the mathematical techniques associated with projectiles in sport. A recent book by Hart and Croft (1988) presents a subset of these techniques for a small selection of events. The aim of the current book is to present a unified collection of the many problems that can be tackled and of all the mathematical techniques that can be employed.

The mathematical foundations of the subject of projectiles were developed from investigations of the motion of bullets and shells. There are some excellent texts describing this research in detail, the best being by McShane, Kelley and Reno (1953). Emphasis here will be on the non-military applications of the behaviour of projectiles in flight, which have received only limited attention.

The purpose of the book is to collect together the various mathematical tools and techniques that will help the reader solve many projectile problems associated with sport or recreation. It begins at an undergraduate level, and emphasises the usefulness of this special topic as a way of teaching mathematical modelling (de Mestre, 1977). A basic knowledge of classical dynamics, calculus, vectors, differential equations and their numerical solution is assumed. At the end of each chapter are exercises, many of which lead on to suitable research projects for honours or masters students. More substantial research topics are indicated in the last chapter, which surveys the application of the mathematical tools and techniques to many recreational and sporting projectile problems.

The first seven chapters of the book encompass material that has been presented to mathematics students at the Royal Military College, Duntroon (1972-1985) and the Australian Defence Force Academy (1986-1988). It analyses the mathematical solution of the equations governing the motion of a projectile during its flight through the air, this motion being generally controlled by gravitational and aerodynamical forces. Although the physical aspects of projectile flight are important, ranging from internal ballistics of a gun, stability, in-flight measurements, through to impact on the ground, they have not been considered in great detail since this has been admirably investigated for military projectiles by Farrar and Leeming (1983), and for other projectiles in some of the many references listed at the end.

In Chapter 1 the classical gravity-only problem is considered with its familiar parabolic trajectory. Next the analysis is extended to include the effects of motion through the air. The main aerodynamic effect for most sporting projectiles is drag, and in Chapter 2 linear drag is investigated initially as this yields a mathematical problem that can be still completely solved by analytical techniques. However, when non-linear drag is considered in Chapter 3, various complexities are introduced that preclude any complete analytical solution, although a semi-analytical technique can be used in special cases. Next follows a chapter on the various numerical techniques that have been developed for the many cases where no progress can be made with analytical techniques.

Chapters 5 and 6 investigate perturbation corrections for two different situations. In the former the ratio of the gravity effect to the drag effect is either very large or very small, and all other forces are neglected. In the latter, gravity and drag are both considered to dominate the flight behaviour, while other forces such as Coriolis, lift, Magnus, variations in gravity and overturning moments are treated as small differential corrections. Spin effects are included in Chapter 7, but the quantitative thrust of the book is concerned essentially with the particle motion of the centre of mass of the projectile, rather than the motion about the centre of mass. Consequently no calculations of precessional rates are attempted, although the complexities of rotational motion are noted. Suitable examples are interspersed throughout the chapters, and in many cases more than one method of solution is given to emphasise the flexibility available.

The main purpose of the previous chapters is to prepare for Chapter 8. This is a representative collection of some of the many applications and occurrences of projectiles in sport, recreation and nature. A very detailed reference list is provided and includes, in particular, survey references for those who wish to "dig deeper" at research level.

I would like to thank my friend Emeritus Professor John Burns for reading the manuscript in draft form and suggesting some much needed changes. Comments by the referees Professor Maurie Brearley and Dr. Vincent Hart were much appreciated. Kaye Schollum and Pauline Hickey did a wonderful job with the typing, while Paul Ballard and Jenny Crook were most meticulous with the diagrams. As always my best friend and wife, Margaret, was a terrific support.

Neville de Mestre

Australian Defence Force Academy

May, 1989.

1. MOTION UNDER GRAVITY ALONE

"Fair pledges of a fruitful tree,

Why do ye fall so fast?"

Robert Herrick (1591 - 1674)

1.1 Gravity

When a small body is projected near a much larger body its trajectory is not straight but curves back towards the larger body. Newton's law of universal gravitation reveals that, when both bodies are spherically symmetric and the small projectile is outside the larger body, the force acting by the larger mass (m_1) on the smaller mass (m_2) is given by

$$\mathbf{F} = -\frac{Gm_1m_2}{r^2}\hat{\mathbf{r}} \tag{1.1}$$

where G is the universal gravitation constant, $\mathbf{r}(= r\hat{\mathbf{r}})$ is the vector from the centre of mass of the larger body to the centre of mass of the smaller body and $\hat{\mathbf{r}}$ is the corresponding unit vector. When the larger body is the Earth $(m_1 = m_e)$ and the small projectile $(m_2 = m)$ is close enough to a point fixed on the Earth's surface, the Earth may be considered to have spherical symmetry with $r \approx r_e$ (the radius of the Earth). Then equation (1.1) is replaced by

$$\mathbf{F} = -\frac{Gm_em}{r_e^2}\hat{\mathbf{j}}$$

where $\hat{\mathbf{j}}$ is a unit vector in the upward vertical direction, which is considered to be constant in both direction and magnitude.

The assumption that $\mathbf{F}$ has a constant direction is called the "flat Earth" assumption and then

$$\mathbf{F} = -mg\hat{\mathbf{j}}$$

$$= m\mathbf{g}$$

where $g = Gm_e/r_e^2$. Since $G = 6.67 \times 10^{-11}$ (M.K.S. units), $m_e = 5.98 \times 10^{24}$(kg) and $r_e = 6.38 \times 10^6$(m) then $g = 9.80$(ms^{-2}), and $\mathbf{g}$ is called the acceleration due to gravity. Its magnitude varies by less than 1% for projectiles within 30 km of the Earth's surface,

and there are similarly small variations for changes in latitude. The force mg is referred
to as the weight of any body of mass m.

1.2 Velocity and Position Vectors

The simplest approximation for a projectile's motion is to consider that
the only force acting on it, after it is launched, is its weight. Then, for motion in free
space, Newton's second law of motion yields

$$m\frac{d^2\mathbf{r}}{dt^2} = m\mathbf{g}$$

that is

$$\frac{d^2\mathbf{r}}{dt^2} = -g\hat{\mathbf{j}} \tag{1.2}$$

where $\mathbf{r}$ is the position vector with respect to a fixed origin on the Earth's surface.

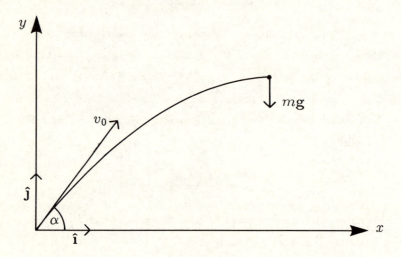

Figure 1.1 The co-ordinate and vector system

If initially $(t = 0)$ the projectile is travelling at speed v_0 at an angle α
to the horizontal the initial velocity vector is

$$\mathbf{v}_0 = v_0 \cos \alpha \; \hat{\mathbf{i}} + v_0 \sin \alpha \; \hat{\mathbf{j}}$$

where $\hat{\mathbf{i}}$ is a unit vector in the horizontal direction forming a right-hand system with $\hat{\mathbf{j}}$ (see Figure 1.1).

When equation (1.2) is integrated with respect to t it yields

$$\mathbf{v} = \mathbf{v}_0 - gt\hat{\mathbf{j}}$$

$$= v_0 \cos \alpha \, \hat{\mathbf{i}} + (v_0 \sin \alpha - gt)\,\hat{\mathbf{j}} \qquad (1.3)$$

where $\mathbf{v} = d\mathbf{r}/dt$ is the velocity vector at any time t. Note that the horizontal component of the velocity of a projectile in free space is a constant.

When equation (1.3) is integrated with respect to time, and the assumption $\mathbf{r} = \mathbf{0}$ when $t = 0$ is made, the position vector is

$$\mathbf{r} = \mathbf{v}_0 t - \frac{1}{2}gt^2\hat{\mathbf{j}}$$

$$= v_0 t \cos \alpha \, \hat{\mathbf{i}} + \left(v_0 t \sin \alpha - \frac{1}{2}gt^2 \right) \hat{\mathbf{j}} \qquad (1.4)$$

If the projectile is at the point (x, y), then $\mathbf{r} = x\hat{\mathbf{i}} + y\hat{\mathbf{j}}$ and the horizontal component of the projectile's displacement is

$$x = \mathbf{r} \cdot \hat{\mathbf{i}} = v_0 t \cos \alpha$$

while the vertical component is

$$y = \mathbf{r} \cdot \hat{\mathbf{j}} = v_0 t \sin \alpha - \frac{1}{2}gt^2$$

These two equations for x and y can be thought of as defining the trajectory in parametric form. When t is eliminated from the two equations they yield

$$y = x \tan \alpha - \frac{gx^2}{2v_0^2} \sec^2 \alpha \qquad (1.5)$$

which can be rearranged to give

$$\left[x - \frac{v_0^2 \sin 2\alpha}{2g} \right]^2 = \frac{2v_0^2 \cos^2 \alpha}{g} \left[\frac{v_0^2 \sin^2 \alpha}{2g} - y \right]$$

This shows that the trajectory is a parabola with its vertex uppermost (see Figure 1.2).

The speed (v) in any position is given by

$$v^2 = \left[\frac{dx}{dt}\right]^2 + \left[\frac{dy}{dt}\right]^2$$

$$= v_0^2 \cos^2 \alpha + (v_0 \sin \alpha - gt)^2$$

$$= v_0^2 \cos^2 \alpha + v_0^2 \sin^2 \alpha - 2v_0 gt \sin \alpha + g^2 t^2$$

$$= v_0^2 - 2gy$$

which could also be obtained by noting that the mechanical energy of the projectile is conserved when only gravity acts.

The angle ψ made by the tangent to the projectile's path at any time t with the horizontal is given by

$$\tan \psi = \frac{dy}{dt} \bigg/ \frac{dx}{dt}$$

$$= \frac{v_0 \sin \alpha - gt}{v_0 \cos \alpha}$$

$$= \tan \alpha - \left[\frac{gt}{v_0}\right] \sec \alpha \qquad (1.6)$$

An alternative form is obtained from equation (1.5) by noting that the slope at any point is

$$\tan \psi = \frac{dy}{dx} = \tan \alpha - \frac{gx \sec^2 \alpha}{v_0^2}$$

1.3 Point of Impact

Impact on the horizontal plane through the projection point occurs when

$$\mathbf{r} \cdot \hat{\mathbf{j}} = 0$$

Thus from equation (1.4),

$$v_0 t \sin \alpha - \frac{1}{2} g t^2 = 0$$

and so

$$t = 0 \quad \text{or} \quad \frac{2v_0 \sin \alpha}{g}$$

Now $t = 0$ corresponds to the projection point, while

$$t_f = \frac{2v_0 \sin \alpha}{g}$$

gives the time of flight to impact.

At the impact point $y = 0$ and so $v^2 = v_0^2$, giving $v = v_0$ since the speed cannot be negative. Also from equation (1.6), the direction at the impact point is given by

$$\tan \psi = \tan \alpha - \frac{g \sec \alpha}{v_0} \left(\frac{2v_0 \sin \alpha}{g} \right)$$

$$= - \tan \alpha$$

Therefore

$$\psi = -\alpha \ \text{ or } \ (\pi - \alpha)$$

but $(\pi - \alpha)$ is irrelevant on physical grounds, since $\psi = \pi - \alpha$ means that the projectile is travelling back along its trajectory. Therefore at impact on the horizontal plane through the projection point the projectile has an angle of depression α, as is obvious from the symmetry of the parabola about its axis.

Now the range (x_f) on the horizontal plane is the x-value when $t = t_f$. Therefore

$$x_f = v_0 \cos \alpha \left(\frac{2v_0 \sin \alpha}{g} \right)$$

$$= \frac{v_0^2 \sin 2\alpha}{g}$$

This result can also be obtained by putting $y = 0$ in equation (1.5). For a given initial speed (v_0) the range is proportional to $\sin 2\alpha$. The maximum range occurs when $\alpha = \pi/4$ radians and has the value v_0^2/g.

1.4 The Vertex of the Trajectory

At the vertex

$$\mathbf{v} \cdot \hat{\mathbf{j}} = 0$$

and so the vertex is reached when

$$t = \frac{v_0 \sin \alpha}{g}$$

$$= \frac{1}{2} t_f$$

Thus at the vertex

$$x = \frac{v_0^2 \sin 2\alpha}{2g}$$

$$y = \frac{v_0^2 \sin^2 \alpha}{2g}$$

$$v = v_0 \cos \alpha$$

$$\psi = 0$$

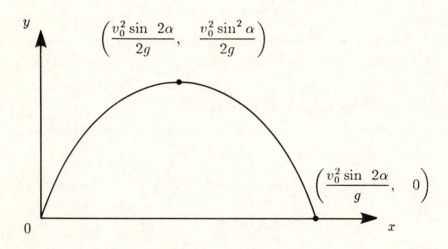

Figure 1.2 The gravity-only trajectory

Example 1.1

    A player hits a baseball into the outfield against a wall 100 metres away. If it leaves the bat at an angle of 45° to the horizontal and strikes the wall 10 metres above the bat-ball contact position, what is the initial speed of the ball?

Solution

Consider equation (1.5)

$$y = x \tan \alpha - \frac{gx^2}{2v_0^2} \sec^2 \alpha$$

On rearranging

$$v_0^2 = \frac{gx^2 \sec^2 \alpha}{2(x \tan \alpha - y)}$$

$$= \frac{9.8 \times 10000 \times 2}{2(100 - 10)}$$

and so

$$v_0 = 33$$

Thus the initial speed is $33 \text{ ms}^{-1} \approx 119 \text{ km h}^{-1}$.

Example 1.2

A cannon mounted on a cliff overlooking the sea can fire a shot at an angle of $30°$ to the horizontal with a muzzle speed of 800 ms^{-1}. If the mouth of the cannon is 100 metres vertically above the base of the cliff, find how far out from the cliff a shot will hit the water.

Solution

(Method 1)

Consider equation (1.4)

$$y = v_0 t \sin \alpha - \frac{1}{2}gt^2$$

Substitution leads to

$$-100 = 400t - 4.9t^2 \qquad (\text{since } \alpha = 30°)$$

which has solutions

$$t = \frac{400 \pm \sqrt{161960}}{9.8}$$

$$\approx 81.88(\text{ the negative answer is irrelevant})$$

Therefore

$$x = v_0 t \cos \alpha$$

$$= 81.88 \times 800 \times \cos \frac{\pi}{6}$$

$$\approx 56728$$

The shot will hit the water approximately 56700 m from the base of the cliff.

(Method 2)

Rewrite equation (1.5) so that the coefficient of x^2 is unity; then

$$x^2 - \frac{v_0^2 \sin 2\alpha}{g} x + \frac{2v_0^2 y \cos^2 \alpha}{g} = 0$$

Substituting the appropriate value yields

$$x^2 - 56556.76x - 9795318.4 = 0$$

The positive root is 56729, and the difference between the results from the two methods is due to the approximate time value used in Method 1.

Example 1.3

On the 1983 tour of Australia by the West Indies' cricket team Joel Garner hit a massive six during one game. The radio commentator said that it was such a big six that the ball went up as high as it went forward. If this was true, at what angle was it hit initially?

Solution

$$\text{Maximum height} = \frac{v_0^2 \sin^2 \alpha}{2g}$$

$$\text{Range} = \frac{v_0^2 \sin 2\alpha}{g}$$

When maximum height equals range it is seen that

$$2 \sin 2\alpha = \sin^2 \alpha$$

which leads to

$$\sin \alpha = 0 \quad \text{or} \quad \tan \alpha = 4$$

Thus $\alpha = 76°$. The answer $0°$ is neglected, since the hit would not be a six.

1.5 Projection from a Different Level

For many projectile problems the projection point is at a different level from the impact point. Suppose that the impact point P is a vertical distance h above the projection point 0 which is selected as the origin. (When P is below 0, h is of course negative). The aim first of all is to determine the horizontal range and time of flight, and then to consider the maximum horizontal range.

When $h > 0$ there will be two points at this height through which the projectile passes. When $h < 0$ however, if the time of flight is restricted to positive values, there is only one point (see Figure 1.3). Now from equation (1.4) since $\mathbf{r} \cdot \hat{\mathbf{j}} = h$ then

$$h = v_0 t \sin \alpha - \frac{1}{2} g t^2$$

The solutions of this quadratic equation in t are

$$t = \frac{v_0 \sin \alpha \pm \sqrt{v_0^2 \sin^2 \alpha - 2gh}}{g}$$

When $v_0 \sin \alpha < \sqrt{2gh}$ the solution is a complex number, indicating that heights $h > v_0^2 \sin^2 \alpha / (2g)$ cannot be reached with these initial conditions. When $v_0 \sin \alpha > \sqrt{2gh}$ there are two cases to consider. For $h > 0$ there are two possible times of flight, with the negative sign associated with the earlier time. For $h < 0$ the expression containing the negative option is neglected.

The horizontal range $(\mathbf{r} \cdot \hat{\mathbf{i}})$ is therefore

$$x = \frac{v_0 \cos \alpha \left\{ v_0 \sin \alpha \pm \sqrt{v_0^2 \sin^2 \alpha - 2gh} \right\}}{g} \tag{1.7}$$

which can also be obtained from equation (1.5) directly.

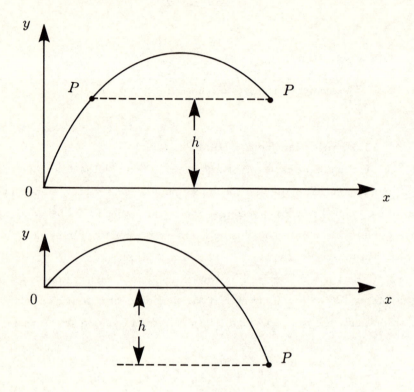

Figure 1.3 Projection through points at height (i) $h > 0$ (ii) $h < 0$

For a given initial speed v_0 the maximum horizontal range is obtained by considering $dx/d\alpha = 0$. It will be left as an exercise for the reader to determine the maximum value of x from equation (1.7). Instead an alternative derivation of the result is given by considering equation (1.5) with $y = h$. Then

$$h = x \tan \alpha - \frac{gx^2}{2v_0^2} \sec^2 \alpha$$

and differentiating with respect to α yields

$$0 = \frac{dx}{d\alpha}\tan\alpha + x\sec^2\alpha - \frac{gx}{v_0^2}\frac{dx}{d\alpha}\sec^2\alpha - \frac{gx^2}{v_0^2}\sec^2\alpha\tan\alpha$$

For a maximum $dx/d\alpha = 0$ and so

$$0 = x\sec^2\alpha - \frac{gx^2}{v_0^2}\sec^2\alpha\tan\alpha$$

which reduces to

$$0 = x\left[1 - \frac{gx}{v_0^2}\tan\alpha\right]$$

Now $x = 0$ does not satisfy equation (1.5) above, while the other solution gives $d^2x/d\alpha^2 < 0$ for α an acute angle. Therefore the maximum value of x satisfies

$$x = \frac{v_0^2}{g}\cot\alpha \tag{1.8}$$

Equations (1.5) with $y = h$ and (1.8) are solved simultaneously to produce the maximum value of x

$$x_m = \frac{v_0\sqrt{v_0^2 - 2gh}}{g} \tag{1.9}$$

and the optimum angle of projection

$$\alpha_m = \arctan\left\{\frac{v_0}{\sqrt{v_0^2 - 2gh}}\right\} \tag{1.10}$$

For $0 \leq h \leq v_0^2/(2g)$ the value of x_m decreases towards zero as h increases, but in contrast the value of α_m increases towards $\pi/2$. When $h > v_0^2/(2g)$ any point at this level cannot be reached by the projectile with this initial speed. For projectiles fired to a level below the projectile point ($h < 0$) the results (1.9) and (1.10) indicate that x_m increases towards infinity and α_m decreases towards zero as $|h|$ increases.

1.6 Inclined Plane

Consider unit vectors $\hat{\mathbf{q}}$ and $\hat{\mathbf{p}}$ along and perpendicular to the inclined plane respectively (Figure 1.4). If β is the angle made by the plane with the horizontal,

$$\hat{\mathbf{q}} = \hat{\mathbf{i}}\cos\beta + \hat{\mathbf{j}}\sin\beta$$

$$\hat{\mathbf{p}} = -\hat{\mathbf{i}}\sin\beta + \hat{\mathbf{j}}\cos\beta$$

When a projectile fired from a point on the inclined plane strikes the plane again, its position vector $\mathbf{r}$ is along the plane and so is perpendicular to $\hat{\mathbf{p}}$.

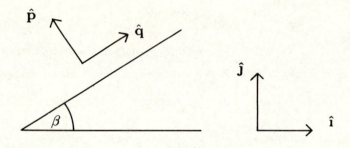

Figure 1.4 The unit vectors for an inclined plane

Thus

$$\mathbf{r} \cdot \hat{\mathbf{p}} = 0$$

and this, with the aid of equation (1.4), leads to an equation for the time of flight. Hence

$$-v_0 t \cos \alpha \sin \beta + \cos \beta \left(v_0 t \sin \alpha - \frac{1}{2} g t^2 \right) = 0$$

$$t \left(-\frac{1}{2} g t \cos \beta + v_0 \sin(\alpha - \beta) \right) = 0$$

and so

$$t = 0 \quad \text{or} \quad \frac{2 v_0 \sin(\alpha - \beta)}{g \cos \beta}$$

Thus the time of flight is

$$t_f = \frac{2 v_0 \sin(\alpha - \beta)}{g \cos \beta}$$

The range on the inclined plane is then

(Method 1)

$$r_f = (\hat{\mathbf{r}} \cdot \hat{\mathbf{q}})_f = v_0 t_f \cos\alpha \cos\beta + \sin\beta \left(v_0 t_f \sin\alpha - \frac{1}{2} g t_f^2 \right)$$

$$= \frac{2v_0^2 \sin(\alpha - \beta) \cos\alpha}{g \cos^2\beta}.$$

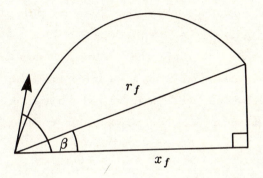

Figure 1.5 Co-ordinates on an inclined plane

(Method 2)

Alternatively, since $\cos\beta = x_f/r_f$ (see Figure 1.5) then

$$r_f = \frac{v_0 t_f \cos\alpha}{\cos\beta}$$

$$= \frac{2v_0^2 \sin(\alpha - \beta) \cos\alpha}{g \cos^2\beta}$$

Rewriting the expression for r_f produces

$$r_f = \frac{v_0^2 \{ \sin(2\alpha - \beta) - \sin\beta \}}{g \cos^2\beta}$$

and so for a given v_0 and β the range is a maximum when

$$\sin(2\alpha - \beta) = 1$$

yielding

$$r_m = \frac{v_0^2}{g(1 + \sin\beta)}$$

Now the angle of projection for maximum range is therefore

$$\alpha = \frac{\pi}{4} + \frac{\beta}{2}$$

$$= \beta + \frac{1}{2}\left\{\frac{\pi}{2} - \beta\right\}$$

Hence the direction for maximum range bisects the angle between the plane and the vertical, and this agrees with the value 45° for the firing from a horizontal plane.

For smaller values of the range, when $\sin(2\alpha - \beta) = \sin\psi$ (say) with $\beta < \psi < \pi/2$, there are then two angles of projection given by $\alpha = \{\pi + 2\beta \pm (\pi - 2\psi)\}/4$; hence two trajectories with the same initial speed, but with directions equally inclined above and below that for maximum range, can be followed to produce a given range.

1.7 Enveloping Parabola

Consider all projectiles fired from a point with initial speed v_0.

The analysis in the previous section shows that at any angle θ to the horizontal the maximum position that could be reached is given by

$$r = \frac{v_0^2}{g(1 + \sin\theta)}$$

where (r, θ) are polar co-ordinates. In Cartesian co-ordinates this becomes

$$y = \frac{v_0^2}{2g} - \frac{gx^2}{2v_0^2} \tag{1.11}$$

which is a parabola called the enveloping parabola, or sometimes called the curve of safety because points outside this curve are safe from the projectiles (Figure 1.6). When the enveloping parabola is rotated about the y-axis the region outside the resulting surface is called the safety zone.

In Example 1.4 below it will be shown that the enveloping parabola is the mathematical envelope of all trajectories fired from the projection point with speed v_0. Hence the theory of envelopes provides an alternative method of obtaining the envelope by considering the trajectory equation (1.5) and differentiating it partially with respect to α keeping x and y fixed. This yields

$$0 = x \sec^2 \alpha - \frac{gx^2}{v_0^2} \sec^2 \alpha \tan \alpha$$

or equivalently, $\tan \alpha = v_0^2/gx$.

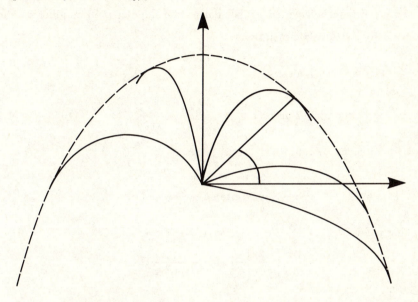

Figure 1.6 The enveloping parabola (dashed curve)

When this is substituted back into equation (1.5) it yields

$$y = x \left(\frac{v_0^2}{gx} \right) - \frac{gx^2}{2v_0^2} \left(1 + \frac{v_0^4}{g^2 x^2} \right)$$

$$= \frac{v_0^2}{2g} - \frac{gx^2}{2v_0^2}$$

Example 1.4

Is the bounding parabola found in equation (1.11) an envelope?

Solution

(Method 1)

This can be checked by calculating the slope of the bounding parabola and the trajectory at any common point.

$$\text{Trajectory}: \qquad y = x \tan \alpha - \frac{gx^2}{2v_0^2} \sec^2 \alpha \qquad (1.5)$$

$$\text{Bounding Parabola}: \qquad y = \frac{v_0^2}{2g} - \frac{gx^2}{2v_0^2} \qquad (1.11)$$

When (x, y) coincides for both

$$\frac{v_0^2}{2g} - \frac{gx^2}{2v_0^2} = x \tan \alpha - \frac{gx^2}{2v_0^2} \sec^2 \alpha$$

Rearranging yields

$$x^2 - \left[\frac{2v_0^2}{g \tan \alpha}\right] x + \frac{v_0^4}{g^2 \tan^2 \alpha} = 0$$

and therefore

$$x = \frac{v_0^2}{g \tan \alpha}$$

$$\text{The slope of the bounding parabola} = -\frac{gx}{v_0^2} = -\cot \alpha$$

$$\text{The slope of the trajectory} = \tan \alpha - \frac{gx}{v_0^2} \sec^2 \alpha$$

$$= \tan \alpha - \frac{\sec^2 \alpha}{\tan \alpha}$$

$$= -\cot \alpha$$

Since the slopes are the same, then the parabola obtained is an envelope. Note that the trajectory touches the enveloping parabola when the projectile has rotated its initial direction through 90°.

(Method 2)

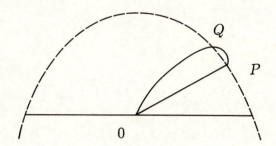

Figure 1.7

Through each point P of the bounding parabola there passes a trajectory (the one which gives maximum range on OP).

Suppose that such a trajectory cuts the boundary parabola at Q (Figure 1.7). But then there would be points between OQ and OP which are outside the bounding parabola and hence greater than the maximum range.

Since this is an obvious contradiction the bounding parabola must be an enveloping parabola.

Example 1.5

Use the parametric equations of the trajectory to obtain the equation of the enveloping parabola.

Solution

The parametric equations of the trajectory are

$$x = v_0 t \cos \alpha$$

$$y = v_0 t \sin \alpha - \frac{1}{2} g t^2$$

The envelope for constant v_0 is governed by the Jacobian determinant

(Courant and John, 1974)

$$\begin{vmatrix} \dfrac{\partial x}{\partial \alpha} & \dfrac{\partial x}{\partial t} \\ \dfrac{\partial y}{\partial \alpha} & \dfrac{\partial y}{\partial t} \end{vmatrix} = 0$$

that is

$$\begin{vmatrix} -v_0 t \sin \alpha & v_0 \cos \alpha \\ v_0 t \cos \alpha & v_0 \sin \alpha - gt \end{vmatrix} = 0$$

This reduces to

$$v_0 t(gt \sin \alpha - v_0) = 0$$

and so

$$t = 0 \quad \text{or} \quad \frac{v_0}{g \sin \alpha}$$

For the non-zero t value the x and y parametric expressions become

$$x = \frac{v_0^2}{g} \cot \alpha,$$

$$y = \frac{v_0^2}{g} - \frac{v_0^2}{2g} \operatorname{cosec}^2 \alpha$$

If α is eliminated between these two equations then as before

$$y = \frac{v_0^2}{2g} - \frac{gx^2}{2v_0^2}$$

Example 1.6

When place kicking a Rugby football a certain player can impart an initial speed $v_0(\text{ms}^{-1})$ to the ball. If the cross bar on the goal-posts is h metres above the field, what is the greatest distance from which he can kick a penalty?

Solution

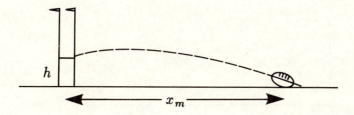

Figure 1.8

Let x_m be the distance from the furthest kicking spot to the goal-posts.

(Method 1)

From the analysis for projection from one level to another, equation (1.9) yields

$$x_m = \frac{v_0 \sqrt{v_0^2 - 2gh}}{g}$$

as the maximum distance in metres from which he can kick a penalty. Note that the angle of projection is given by equation (1.10).

(Method 2)

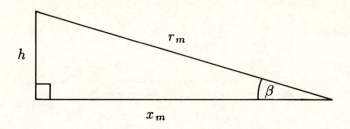

Figure 1.9

The problem can be treated as projectile motion up a plane of inclination β where

$$\sin \beta = \frac{h}{r_m}$$

The maximum range on this plane is

$$r_m = \frac{v_0^2}{g(1 + \sin \beta)}$$

$$= \frac{v_0^2 r_m}{g(r_m + h)}$$

Therefore

$$r_m = \frac{v_0^2 - gh}{g}$$

and so

$$x_m = \sqrt{r_m^2 - h^2}$$

$$= \frac{v_0 \sqrt{v_0^2 - 2gh}}{g}$$

(Method 3)

The problem can also make use of the envelope results. For maximum range to the goal-posts the envelope passes just over the crossbar. Therefore from equation (1.11)

$$h = \frac{v_0^2}{2g} - \frac{gx_m^2}{2v_0^2}$$

and so

$$x_m = \frac{v_0\sqrt{v_0^2 - 2gh}}{g}$$

1.8 Dimensionless Equations

Throughout this chapter the expression v_0/g is associated with the time of flight expressions, while v_0^2/g appears in all expressions for x, y and the range.

Physical variables such as time, length, velocity, acceleration and many others should be expressed in appropriately measured units (second or hour, metre or kilometre, and so on). When a physical (or dimensional) variable is divided by a known measurement in the same units it is then a non-dimensional variable.

For example, if a car is travelling at 80 km h^{-1} the air particles in the vicinity of the car may have dimensional speeds ranging from 0 to 80 km h^{-1}. Suppose that a non-dimensional variable V is introduced equal to air particle speed divided by car speed, then the air particles will have non-dimensional speeds ranging from 0 to 1. The car speed 80 km h^{-1} is called a representative speed for the problem.

The important results of this chapter will now be rewritten in non-dimensional form. To do this a representative time and a representative length are needed.

Since v_0 has the dimensions of length/time and g has the dimensions of length/(time)^2 it is no surprise to observe that a representative time is v_0/g and a representative length is v_0^2/g. Therefore, non-dimensional variables are defined by

$$T = \frac{gt}{v_0}, \quad X = \frac{gx}{v_0^2}, \quad Y = \frac{gy}{v_0^2} \tag{1.12}$$

Since angles are dimensionless the dimensionless forms of equations (1.4) - (1.7) and

(1.9) - (1.11) become

$$\mathbf{R} = T \cos \alpha \, \hat{\imath} + \left(T \sin \alpha - \frac{1}{2}T^2 \right) \hat{\jmath} \qquad (1.4*)$$

$$Y = X \tan \alpha - \frac{1}{2}X^2 \sec^2 \alpha \qquad (1.5*)$$

$$\tan \psi = \tan \alpha - T \sec \alpha \qquad (1.6*)$$

$$X = \cos \alpha \left(\sin \alpha \pm \sqrt{\sin^2 \alpha - 2H} \right) \qquad (1.7*)$$

$$X_m = \sqrt{1 - 2H} \qquad (1.9*)$$

$$\alpha_m = \arctan \left\{ \frac{1}{\sqrt{1 - 2H}} \right\} \qquad (1.10*)$$

and

$$Y = \frac{1}{2} - \frac{1}{2}X^2 \qquad (1.11*)$$

where

$$H = \frac{gh}{v_0^2}$$

These are clearly simpler in form than the corresponding dimensional equations. The use of dimensionless variables will greatly simplify the analysis of later chapters also. The dimensional results can be easily recovered at the end of the analysis through the transformations (1.12).

1.9 Exercises

1. Obtain expressions for x, y, v, t as functions of the angle ψ made by the tangent to the projectile's path with the horizontal.

2. Show that for a given range and muzzle speed there are in general two values of α possible (that is, two trajectories), provided that $v_0^2 > gx_f$. Comment on $v_0^2 \le gx_f$.

3. Is it possible to throw a stone from the top of the pyramid of Cheops so that it strikes the ground beyond the base? The height of the pyramid is 137.2 m, the length of each side of its square base is 227.5 m and the speed with which a person can throw a stone is 24 ms^{-1}.

4. A fort is on top of a cliff h metres directly above the ocean. Approaching the fort is a ship whose guns have the same muzzle velocity v_0 as the guns at the fort. Neglecting all effects other than gravity find over what range the ship can be fired on, from the fort, without being able to effectively return the fire. If gh is small compared with v_0^2 show that this distance is approximately double the height of the cliff.

5. Stones are thrown with speed v_0 so as to clear a wall of height h at a distance a from the point of projection. Show that the distance immediately beyond the wall within which no stone can land is

$$\frac{ah\left(v_0^2 - gh - \sqrt{v_0^4 - 2ghv_0^2 - g^2a^2}\right)}{g\left(a^2 + h^2\right)}$$

6. Determine the closest distance that a place-kicker in rugby can score from if his kicking angles are limited by $0 \le \alpha \le \alpha_0$.

7. Use equation (1.7) directly to obtain the values for x_m and α_m given in equations (1.9) and (1.10).

8. A particle is projected at an angle α to the horizontal up a plane of inclination β. If the particle strikes the plane at right angles prove that

$$\tan \alpha = \cot \beta + 2 \tan \beta$$

9. Show that for maximum range, the angle of projection at speed v_0 from a height h is given by

$$\alpha_m = \arcsin\left[\left\{2\left[1 + \frac{gh}{v_0^2}\right]\right\}^{-\frac{1}{2}}\right]$$

10. For a shell fired with speed v_0 and angle α relative to a platform to which a gun is fixed, determine the position vector of the shell at any time t when the platform has a constant speed U in the horizontal direction of the shell's motion. Find the maximum height and horizontal range of the trajectory.

11. Rewrite expressions for the range and time of flight on a plane of inclination β in non-dimensional form.

2. MOTION IN A LINEAR RESISTING MEDIUM

"Even horizontal motion, which if no impediment were offered
would be uniform and constant, is altered by the
resistance of the air and finally ceases, and here again the
less dense the body, the quicker the process."

Galileo Galilei (1564–1642)

2.1 Velocity and Position Vectors

The formulae derived in Chapter 1 relate strictly to a projectile travelling
under the influence of constant gravity in a vacuum. When the projectile moves
through any fluid medium (gas or liquid) other forces are present due to the slowing-
down influence of that medium's particles. The sum of these forces in a direction
opposite to the projectile's velocity is called the drag force and for many non-spinning
projectiles it is the main effect of the medium. A more detailed discussion of drag will
be postponed until the beginning of Chapter 4.

Experiments show that this drag force is usually related in a non-linear
way to the velocity of the projectile. When a projectile is moving at moderate or high
speeds the non-linear drag force can be approximated by using different powers of the
speed over different velocity ranges.

For projectiles moving through air at very low speeds, or for motion
through other fluids where the Reynolds number (see Chapter 7) is small, the drag
can often be assumed to be directly proportional to the speed. This linear model shows
the effect of the inclusion of drag without the mathematics becoming too complicated
at first. This approximation will be pursued here and other drag approximations will
be considered in the next chapter.

The resisting force exerted on a projectile by the air always acts in the
direction tangential to the path of the projectile and so continually changes in direction
as the trajectory curves back towards the earth. In general, this variation in direction of
the drag causes difficulties in the mathematical solution because it leads to a system of
coupled non-linear differential equations. The particular advantage of the assumption

that the drag is proportional to the speed is that this is the only case in which the
governing equations are linear and uncoupled.

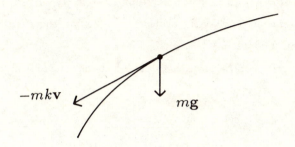

Figure 2.1 Gravity and linear drag

If the projectile has position vector $\mathbf{r}$ at time t, Newton's law of motion
yields

$$\frac{d^2\mathbf{r}}{dt^2} = -g\hat{\mathbf{j}} - k\mathbf{v} \tag{2.1}$$

where k is a constant called the resistance coefficient per unit mass (see Figure 2.1).
Note that $k\mathbf{v}$ has the dimensions of acceleration and so k must have the dimensions
of $(\text{time})^{-1}$. If dimensionless variables are introduced through equation (1.12) then

$$\frac{d\mathbf{V}}{dT} + \epsilon\mathbf{V} = -\hat{\mathbf{j}}$$

with $\mathbf{v} = v_0\mathbf{V}$, where $\epsilon = kv_0/g$ is a nondimensional constant. Now ϵ may be interpreted as the ratio of the initial drag to the weight. Thus small ϵ indicates that the
drag effect is much smaller than the gravitational effect, while large ϵ indicates that
the drag force dominates the gravitational force. In the latter case the motion of the
projectile is almost rectilinear from the beginning of the trajectory until the drag force
slows it down to a position where the gravitational force starts to have an appreciable
effect.

This linear non-homogeneous vector differential equation can easily be solved to obtain

$$\mathbf{V} = \hat{\mathbf{v}}_0 e^{-\epsilon T} + \frac{1}{\epsilon}\left(e^{-\epsilon T} - 1\right)\hat{\mathbf{j}} \qquad (2.2)$$

where $\hat{\mathbf{v}}_0 = [\cos\alpha,\ \sin\alpha]$ is the initial dimensionless velocity of projection. Since $\epsilon T = kt$, the horizontal component of the projectile's velocity is therefore $v_0 \cos\alpha\ e^{-kt}$, and so is not constant unless $k = 0$ ($\epsilon = 0$).

The direction of the projectile's motion therefore makes an angle ψ with the horizontal, where ψ is given by

$$\tan\psi = \frac{\epsilon\sin\alpha + \left(1 - e^{\epsilon T}\right)}{\epsilon\cos\alpha}$$

$$= \tan\alpha + \frac{\sec\alpha(1 - e^{\epsilon T})}{\epsilon} \qquad (2.3)$$

Use of L'Hôpital's rule shows that this approaches the free space result (1.6*) as ϵ approaches zero.

When equation (2.2) is integrated with respect to T, and the initial condition $\mathbf{R} = \mathbf{0}$ is used, then

$$\mathbf{R} = \frac{\hat{\mathbf{v}}_0}{\epsilon}\left(1 - e^{-\epsilon T}\right) + \frac{1}{\epsilon^2}\left(1 - \epsilon T - e^{-\epsilon T}\right)\hat{\mathbf{j}} \qquad (2.4)$$

The equation of the trajectory is obtained by eliminating T from the components of $\mathbf{R}$ in the $\hat{\mathbf{i}}$ and $\hat{\mathbf{j}}$ directions. Thus

$$X = \frac{\cos\alpha}{\epsilon}\left(1 - e^{-\epsilon T}\right),$$

$$Y = \frac{\sin\alpha}{\epsilon}\left(1 - e^{-\epsilon T}\right) + \frac{1}{\epsilon^2}\left(1 - \epsilon T - e^{-\epsilon T}\right)$$

and so

$$Y = X\tan\alpha + \left(\epsilon X\sec\alpha + \ln(1 - \epsilon X\sec\alpha)\right)/\epsilon^2 \qquad (2.5)$$

For $T \geq 0$ the value of X is bounded and can never be greater than $\cos\alpha/\epsilon$. This property is related to the terminal velocity that the projectile reaches under this drag law. From equation (2.1) it is seen that when $\mathbf{v} = -(g/k)\hat{\mathbf{j}}$, the acceleration $d^2\mathbf{r}/dt^2$ is zero. This means that if the particle is projected vertically downwards with speed

g/k (so that its initial velocity is $\mathbf{v}_0 = -(g/k)\hat{\mathbf{j}}$) then it will have zero acceleration and will always have this velocity. For any other value of $\mathbf{v}_0$, equation (2.2) shows that $\mathbf{v} \to -(g/k)\hat{\mathbf{j}}$ as $t \to \infty$ (that is, $\mathbf{V} \to -(1/\epsilon)\hat{\mathbf{j}}$ as $T \to \infty$) and this velocity $\mathbf{v}$ is called the terminal velocity of the projectile. Since $e^{-\epsilon T} \to 0$ very rapidly as T increases (for example, $e^{-\epsilon T} < 10^{-6}$ when $\epsilon T = 14$), the velocity of the particle is in practice indistinguishably close to the terminal velocity after finite (and often quite small) times. Further discussion of terminal speeds will be postponed until the next chapter when more general drag laws are considered.

A graphical comparison of equation (2.5) with the analogous no-drag equation (1.5*) shows that the presence of drag removes the symmetry of the flight path (see Figure 2.2). The projectile comes down in a path steeper than the one in which it ascended.

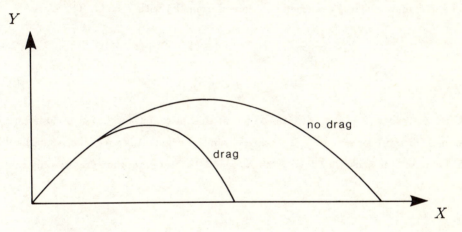

Figure 2.2 Comparison of trajectory equations (1.5*) and (2.5)

Formulae generated in the next three sections will enable the reader to see that the presence of drag in the form proposed in this model results in a reduction in the maximum height, a reduction in the horizontal range, and an increase in the angle of impact (see 2.8, Exercise 3). These tendencies persist when non-linear models for the drag effect are also considered.

2.2 Maximum Height

The vertex of the trajectory occurs when

$$\mathbf{V}.\hat{\mathbf{j}} = 0$$

Therefore from equation (2.2)

$$e^{-\epsilon T_z} \sin\alpha + \frac{1}{\epsilon}\left(e^{-\epsilon T_z} - 1\right) = 0$$

where T_z denotes the time to reach the maximum height or zenith of the trajectory. Rearranging this yields

$$T_z = \frac{1}{\epsilon}\ln(1 + \epsilon\sin\alpha) \tag{2.6}$$

The maximum height Y_z reached is obtained from $\mathbf{R}.\hat{\mathbf{j}}$ with $T = T_z$. Therefore

$$Y_z = \frac{\sin\alpha}{\epsilon}\left(1 - e^{-\epsilon T_z}\right) + \frac{1}{\epsilon^2}\left(1 - \epsilon T_z - e^{-\epsilon T_z}\right)$$

$$= \frac{\sin\alpha}{\epsilon} - \frac{1}{\epsilon^2}\ln(1 + \epsilon\sin\alpha) \tag{2.7}$$

Using equations (2.4) and (2.6) it is seen that this maximum height occurs at a dimensionless distance $\sin\alpha\cos\alpha \,/\, (1+\epsilon\sin\alpha)$ from the origin. It is a useful check (Chapter 2, Exercise 1) to show that T_z, Y_z and X_z reduce to the appropriate gravity-only values as $\epsilon \to 0$.

2.3 Impact Time to the Horizontal Plane

This occurs when $\mathbf{R}.\hat{\mathbf{j}} = 0$. Thus from equation (2.4)

$$\frac{\sin\alpha}{\epsilon}\left(1 - e^{-\epsilon T_f}\right) + \frac{1}{\epsilon^2}\left(1 - \epsilon T_f - e^{-\epsilon T_f}\right) = 0$$

where T_f denotes the dimensionless time of flight. This is a transcendental equation for T_f and is difficult to solve analytically. It can be rewritten as

$$1 - e^{-\epsilon T_f} = \frac{\epsilon T_f}{\epsilon\sin\alpha + 1} \tag{2.8}$$

and this form is useful for finding the solution graphically as indicated by the sketch in Figure 2.3.

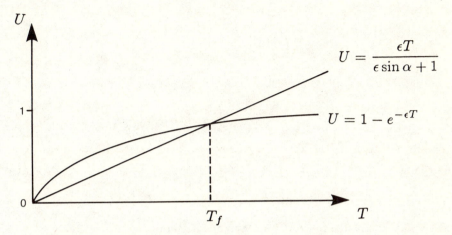

Figure 2.3 Sketch of the graphical solution of equation (2.*)

For many problems v_0 is small enough so that kt (or ϵT) is small compared with unity. Equation (2.8) can then be solved approximately by expanding $e^{-\epsilon T_f}$ in its Maclaurin series; thus

$$\epsilon T_f - \frac{\epsilon^2 T_f^2}{2} + \cdots = \frac{\epsilon T_f}{1 + \epsilon \sin \alpha}$$

and since $T_f \neq 0$ this produces

$$T_f \approx \frac{2 \sin \alpha}{1 + \epsilon \sin \alpha}$$

However ϵT_f was assumed large, and therefore for consistency $2\epsilon \sin \alpha / (1 + \epsilon \sin \alpha) \ll 1$ which is equivalent to $\epsilon \ll 1$. Thus

$$T_f = 2 \sin \alpha + 0(\epsilon)$$

for small ϵ, which says that the leading term is the no-drag term. The first-order drag correction can be made to the time of flight by writing formally

$$T_f = T_0 + \epsilon T_1 + 0[\epsilon^2]$$

and substituting this into equation (2.8). After equating terms of like order in powers of ϵ, this yields

$$T_0 = 2\sin\alpha$$

$$T_1 = -\frac{2}{3}\sin^2\alpha$$

and so for $kv_0 \ll g$ the time of flight is given by

$$t_f = \frac{2v_0\sin\alpha}{g} - \frac{2kv_0^2}{3g^2}\sin^2\alpha + 0\left[\frac{k^2v_0^2}{g^2}\right]$$

The time of flight in this case is therefore less than that in the no-drag case derived in Chapter 1.

2.4 Range on the Horizontal Plane

The range on the horizontal plane is obtained from $\mathbf{R}\cdot\hat{\imath}$ when $T = T_f$. Thus

$$X_f = \frac{\cos\alpha\left(1 - e^{-\epsilon T_f}\right)}{\epsilon}$$

$$= \frac{T_f\cos\alpha}{1 + \epsilon\sin\alpha}$$

using equation (2.8). When T_f has been determined this form is very useful for calculating the range.

For small ϵT_f the approximate range is given by

$$X_f = \frac{\cos\alpha\left(2\sin\alpha - \frac{2}{3}\epsilon\sin^2\alpha + 0(\epsilon^2)\right)}{1 + \epsilon\sin\alpha}$$

$$= \cos\alpha\left(2\sin\alpha - \frac{8}{3}\epsilon\sin^2\alpha + 0(\epsilon^2)\right)$$

thus

$$x_f = \frac{v_0^2\sin 2\alpha}{g}\left[1 - \frac{4kv_0}{3g}\sin\alpha + 0\left[\frac{k^2v_0^2}{g^2}\right]\right]$$

Now $v_0^2\sin 2\alpha/g$ is the no-drag result while $-4kv_0^3\sin 2\alpha\sin\alpha/(3g^2)$ gives the approximate correction for small kt_f. For this approximation the angle of projection for maximum range is given by

$$\sin\alpha_m = \frac{1}{\sqrt{2}} - \frac{1}{6}\epsilon + 0(\epsilon^2)$$

that is

$$\alpha_m = \frac{\pi}{4} - \frac{\sqrt{2}}{6}\epsilon + 0(\epsilon^2)$$

Notice that the presence of drag leads to a smaller angle of projection for maximum range. Therefore

$$\cos \alpha_m = \frac{1}{\sqrt{2}} + \frac{1}{6}\epsilon + 0(\epsilon^2)$$

and so

$$X_m = 1 - \frac{2\sqrt{2}}{3}\epsilon + 0(\epsilon^2)$$

This indicates that the maximum range is reduced by a quantity $2\sqrt{2}kv_0^3/(3g^2)$ when there is a small amount of drag present.

2.5 Envelopes

By considering the appropriate envelope the danger zone bounded by the curve of safety is defined. Within this zone lie all points that can be reached by a projectile fired from a fixed point with muzzle speed v_0. The equation of the envelope is obtained by eliminating α between equation (2.5) and the derivative of this equation with respect to α put equal to zero. The latter yields

$$0 = X \sec^2 \alpha + \frac{X \sec \tan \alpha}{\epsilon}\left[1 - \frac{1}{1 - \epsilon X \sec \alpha}\right]$$

which reduces to

$$X = \frac{\cos \alpha}{\sin \alpha + \epsilon} \tag{2.9}$$

Substitution into equation (2.5) yields

$$Y = \frac{1 + \epsilon \sin \alpha}{\epsilon(\sin \alpha + \epsilon)} + \frac{1}{\epsilon^2} \ln \left(\frac{\sin \alpha}{\sin \alpha + \epsilon}\right) \tag{2.10}$$

Equations (2.9) and (2.10) are parametric equations for the envelope. Elimination of α produces the envelope equation in the form $Y = f(X)$. This will be done in the next section.

As a check it should be noted that for $\alpha = \pi/2$ the defined envelope point is seen to correspond to the point of maximum height as given by equation (2.7) for a projectile with this muzzle angle.

Example 2.1

Obtain the parametric equations for the envelope of a projectile under a linear drag law using equation (2.4) directly (Murphy, 1972).

Solution

Now equation (2.4) has components

$$X = \frac{\cos \alpha}{\epsilon} \left(1 - e^{-\epsilon T} \right)$$

$$Y = \frac{\sin \alpha}{\epsilon} \left(1 - e^{-\epsilon T} \right) + \frac{1}{\epsilon^2} \left(1 - \epsilon T - e^{-\epsilon T} \right)$$

From the discussion of envelopes in Chapter 1

$$\begin{vmatrix} \dfrac{\partial X}{\partial \alpha} & \dfrac{\partial X}{\partial T} \\[2mm] \dfrac{\partial Y}{\partial \alpha} & \dfrac{\partial Y}{\partial T} \end{vmatrix} = 0$$

and so

$$\begin{vmatrix} -\sin \alpha \left(1 - e^{-\epsilon T} \right) / \epsilon & \cos \alpha\, e^{-\epsilon T} \\[2mm] \cos \alpha \left(1 - e^{-\epsilon T} \right) / \epsilon & \sin \alpha\, e^{-\epsilon T} + \left(e^{-\epsilon T} - 1 \right) / \epsilon \end{vmatrix} = 0$$

which simplifies to

$$e^{-\epsilon T} = \frac{\sin \alpha}{\epsilon + \sin \alpha}$$

When this is substituted into the X and Y expressions above, the results are

$$X = \frac{\cos \alpha}{\epsilon + \sin \alpha}$$

$$Y = \frac{1 + \epsilon \sin \alpha}{\epsilon(\epsilon + \sin \alpha)} + \frac{1}{\epsilon^2} \ln \left(\frac{\sin \alpha}{\epsilon + \sin \alpha} \right)$$

as before.

2.6 Inclined Planes

For impact on a plane of inclination β a similar analysis to that given in Section 2.4 produces the time of flight (when $\mathbf{R} \cdot \hat{\mathbf{p}} = 0$) and the range ($\mathbf{R} \cdot \hat{\mathbf{q}}$ with $T = T_f$), and this is left as an exercise for the reader at the end of the chapter.

The results of Section 2.5 are useful when investigating maximum range on an inclined plane. Since β is the angle of the plane the co-ordinates (X_m, Y_m) of the point of impact at maximum range satisfy

$$\tan \beta = \frac{Y_m}{X_m}$$

Therefore from equations (2.9) and (2.10) the angle of projection α_m for maximum range on a plane of inclination β is the solution of

$$\tan \beta = \frac{1 + \epsilon \sin \alpha_m}{\epsilon \cos \alpha_m} + \frac{(\epsilon + \sin \alpha_m)}{\epsilon^2 \cos \alpha_m} \ln \left(\frac{\sin \alpha_m}{\sin \alpha_m + \epsilon} \right)$$

With the transformation $W_m = 1 + \epsilon \operatorname{cosec} \alpha_m$ this becomes

$$W_m - 1 + \epsilon^2 - W_m \ln W_m - \epsilon \tan \beta \sqrt{(W_m - 1)^2 - \epsilon^2} = 0 \qquad (2.11)$$

This equation can be solved graphically or iteratively by Newton's successive approximations using a computer or a calculator (Kreyszig, 1983). Once W_m is known then

$$\alpha_m = \arcsin \left[\frac{\epsilon}{W_m - 1} \right] \qquad (2.12)$$

and the maximum range can be obtained from

$$X_m = \frac{\cos \alpha_m}{\epsilon + \sin \alpha_m}$$

$$= \frac{\cot \alpha_m}{W_m} \qquad (2.13)$$

In Figures 2.4 and 2.5 the maximum range and optimum projection angle are plotted against ϵ for impact on the horizontal plane ($\beta = 0$).

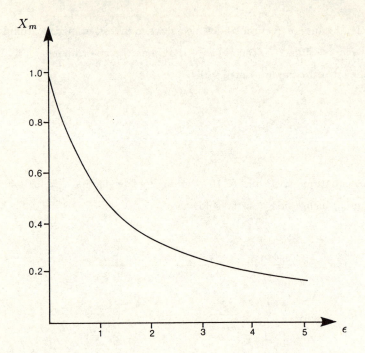

Figure 2.4 X_m versus ϵ when $\beta = 0$

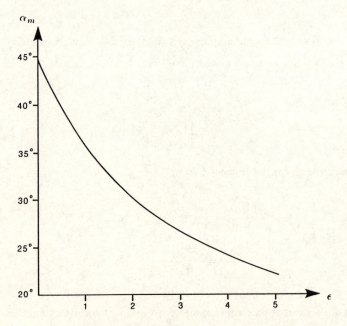

Figure 2.5 α_m versus ϵ when $\beta = 0$

Example 2.2

Determine α_m, X_m and x_m when $\epsilon = 1$ and $\beta = 0$.

Solution

When $\epsilon = 1$ and $\beta = 0$ then

$$W_m - W_m \ln W_m = 0$$

The relevant solution is $W_m = e$, and the optimum angle of projection from equation (2.12) is

$$\alpha_m = \arcsin \left[\frac{1}{e-1} \right]$$

$$\approx 35.59°$$

From equation (2.13)

$$X_m \approx \frac{\cot 35.59°}{e}$$

$$\approx 0.51$$

and the maximum range is given approximately by

$$x_m \approx \frac{0.51 v_0^2}{g}$$

Thus when the initial drag and gravity are equal in magnitude the linear drag law predicts that the maximum range is approximately half the no-drag result.

2.7 Curve of Safety

Equations (2.9) and (2.10) give the co-ordinates of the point where the trajectory touches the curve of safety. If equation (2.9) is solved for α it is seen that

$$\sin \alpha = \frac{\sqrt{1 + X^2 - \epsilon^2 X^2} - \epsilon X^2}{(1 + X^2)}$$

Substituting this into equation (2.10) produces

$$Y = \begin{cases} \dfrac{\sqrt{1 + X^2 - \epsilon^2 X^2}}{\epsilon} + \dfrac{1}{\epsilon^2} \ln \left(\dfrac{1 - \epsilon\sqrt{1 + X^2 - \epsilon^2 X^2}}{1 - \epsilon^2} \right) & (\epsilon \neq 1) \\[4ex] 1 + \ln \left| \dfrac{1}{2} \left(1 - X^2 \right) \right| & (\epsilon = 1) \end{cases}$$

which are the Cartesian forms of the curve of safety. These are illustrated in Figure 2.6.

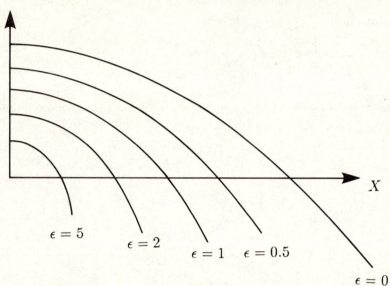

Figure 2.6 Curves of safety for different ϵ values

Example 2.3

Show that as $\epsilon \to 0$ the curve of safety is approximately a parabola. What are the dimensional and non-dimensional forms of this parabola? What is the next-order correction term for ϵ small and $\epsilon X \ll 1$?

Solution

For $\epsilon \neq 1$ the cartesian form of the curve of safety is

$$Y = \frac{\sqrt{1 + X^2 - \epsilon^2 X^2}}{\epsilon} + \frac{1}{\epsilon^2} \ln \left(\frac{1 - \epsilon\sqrt{1 + X^2 - \epsilon^2 X^2}}{1 - \epsilon^2} \right)$$

As $\epsilon \to 0$ with $\epsilon X \ll 1$

$$\frac{\sqrt{1 + X^2 - \epsilon^2 X^2}}{\epsilon} = \frac{\sqrt{1 + X^2}}{\epsilon} - \frac{\epsilon X^2}{2\sqrt{1 + X^2}} + 0(\epsilon^3)$$

and

$$\frac{1}{\epsilon^2}\ln\left(\frac{1-\epsilon\sqrt{1+X^2}-\epsilon^2 X^2}{1-\epsilon^2}\right) = \frac{-\sqrt{1+X^2}}{\epsilon} + \frac{1}{2}(1-X^2)$$

$$+\epsilon\left\{\frac{X^2}{2\sqrt{1+X^2}}-\frac{1}{3}(1+X^2)^{\frac{3}{2}}\right\}+0(\epsilon^2)$$

Therefore

$$Y = \frac{1}{2}(1-X^2) - \frac{\epsilon}{3}\left(1+X^2\right)^{\frac{3}{2}} + 0(\epsilon^2)$$

or

$$Y \approx \frac{1}{2}(1-X^2)$$

which is approximately a parabola in non-dimensional form. Reverting to dimensional form it is seen that for $\epsilon = 0$ the trajectory is

$$\frac{gy}{v_0^2} = \frac{1}{2}\left[1 - \frac{g^2 x^2}{v_0^4}\right]$$

that is

$$y = \frac{v_0^2}{2g} - \frac{gx^2}{2v_0^2}$$

as was obtained in Chapter 1.

The next order non-dimensional correction is clearly of order ϵ with a coefficient $-\frac{1}{3}(1+X^2)^{\frac{3}{2}}$. Thus for small kv_0/g

$$y = \frac{v_0^2}{2g} - \frac{gx^2}{2v_0^2} - \frac{k}{3g^2 v_0^3}\left(v_0^4 + g^2 x^2\right)^{\frac{3}{2}} + 0\left(\frac{k^2 v_0^2}{g^2}\right)$$

2.8 Exercises

1. Check that equations (2.2), (2.3) and (2.4) reduce to the no-drag results as $\epsilon \to 0$. Is this also true for the equation of the trajectory given by (2.5)?

2. Show that the two component equations for (2.1) in the $\hat{\imath}$ and $\hat{\jmath}$ directions are an uncoupled system of differential equations. With appropriate

initial conditions for $x, dx/dt$ and dy/dt solve these equations and verify the corresponding solutions obtained by the vector approach.

3. Sketch the form of the trajectories given by equations (2.5) and (1.5) for $v_0 = 70$, $\alpha = \pi/6$ and $k = 1/7$.

4. Obtain the speed of a projectile as a function of t when it is acted upon by gravity and a drag proportional to its speed. Does v^2 obtained from the law of conservation of mechanical energy agree with this expression? Explain.

5. Obtain expressions for the time of flight and range of a projectile under a linear drag law impacting on a plane of inclination β.

6. Determine approximately the curve of safety for ϵ large with X small such that $\epsilon X = 0(1)$. Describe it.

7. When a particle moves through a viscous fluid (oil, glucose, etc.) its drag is proportional to its speed. For a particle projected horizontally in one of these fluids, whose viscosity is such that the drag force is much greater than the weight of the particle, determine its approximate trajectory.

8. Show that when the trajectory touches the curve of safety the velocity is perpendicular to its initial direction. What does this say about the angle of impact on an inclined plane when the range is a maximum?

9. Solve the W_m equation (2.11) for $\epsilon = 0.5$ and $\beta = \pi/4$, and use the solution to determine α_m and X_m.

3. MOTION IN A NON-LINEAR RESISTING MEDIUM

"O, he flies through the air with the greatest of ease,

This daring young man on the flying trapeze."

George Leybourne (? - 1884)

3.1 Non-linear Drag

In many problems involving the motion of a projectile the drag force is not linear with respect to speed but assumes a more complicated form. Many investigations have recognised that if $\rho(y)$ is the density of air at height y, A is the cross-sectional area of the projectile and $\mathbf{v}$ is the velocity of the projectile, then the combination $\rho A v^2$ has the dimensions of force. Therefore, the drag force $\mathbf{D}$ can be expressed as

$$\mathbf{D} = -\frac{1}{2}\rho A v^2 C_D \hat{\mathbf{v}} \qquad (3.1)$$

where C_D is known as the drag coefficient. This coefficient is non-dimensional and for a given projectile it may be a function of the projectile's speed, the angle of attack (yaw) of the projectile to its path and the air temperature at different altitudes. It will also vary from one projectile to another depending on the amount of streamlining, the surface roughness and the attachment of drag-enhancing equipment such as parachutes. Typical values of C_D range between 0 and 2.

However, in some contexts ρ is constant and the combination AC_D is either a dimensional constant or proportional to some power of v. It is therefore worthwhile to consider the general case of a power-law resisting medium where the magnitude of the drag is proportional to v^n (with n a positive constant). When $n = 1$ the results of Chapter 2 should be recovered. At the end of this chapter other non-linear drag expressions will be considered.

3.2 Cartesian Equations for Power-law Drag

Newton's law of motion for a non-spinning projectile moving under gravity in this medium is

$$\frac{d^2\mathbf{r}}{dt^2} = -g\hat{\mathbf{j}} - kv^n\hat{\mathbf{v}} \tag{3.2}$$

When $n = 2$ it is seen that $k = \rho A C_D/(2m)$ which has the dimensions of $(\text{length})^{-1}$.

The terminal velocity occurs when $d^2\mathbf{r}/dt^2 = \mathbf{0}$, and so the terminal speed is therefore $(g/k)^{\frac{1}{n}}$ occurring when $\hat{\mathbf{v}} = -\hat{\mathbf{j}}$. For a person free-falling in air it is usual to assume $n = 2$ and then the terminal speed depends on $k^{-\frac{1}{2}}$. The constant k can be adjusted for a given projectile by changing the horizontal cross-sectional area of the falling body, so altering its terminal speed. This property is frequently applied in sky-diving. Terminal speeds of the order of 225 km h $^{-1}$ are common in this sport.

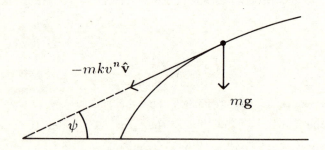

Figure 3.1 Force diagram for projectile influenced by gravity and non-
linear drag

If equation (3.2) is separated into its $\hat{\mathbf{i}}$ and $\hat{\mathbf{j}}$ components it yields

$$\frac{d^2x}{dt^2} = -k\left\{\left[\frac{dx}{dt}\right]^2 + \left[\frac{dy}{dt}\right]^2\right\}^{n/2}\cos\psi$$

$$\frac{d^2y}{dt^2} = -g - k\left\{\left[\frac{dx}{dt}\right]^2 + \left[\frac{dy}{dt}\right]^2\right\}^{n/2}\sin\psi$$

where ψ is the angle that the tangent makes with the horizontal as shown in Figure

3.1. But

$$\cos\psi = \frac{dx}{dt}\left\{\left[\frac{dx}{dt}\right]^2 + \left[\frac{dy}{dt}\right]^2\right\}^{-\frac{1}{2}}, \quad \sin\psi = \frac{dy}{dt}\left\{\left[\frac{dx}{dt}\right]^2 + \left[\frac{dy}{dt}\right]^2\right\}^{-\frac{1}{2}}$$

and so

$$\left.\begin{aligned}\frac{d^2x}{dt^2} &= -k\frac{dx}{dt}\left\{\left[\frac{dx}{dt}\right]^2 + \left[\frac{dy}{dt}\right]^2\right\}^{(n-1)/2}\\[2mm]\frac{d^2y}{dt^2} &= -k\frac{dy}{dt}\left\{\left[\frac{dx}{dt}\right]^2 + \left[\frac{dy}{dt}\right]^2\right\}^{(n-1)/2} - g\end{aligned}\right\} \qquad (3.3)$$

This is a system of coupled highly non-linear differential equations, which only uncouple when $n = 1$ (that is, drag proportional to v). These equations have not yet been completely solved by analytical methods alone. However Parker (1977) wrote them in non-dimensional form and then obtained approximate solutions for low angles of trajectory by assuming $V^2/U^2 \ll 1$.

They can be partially solved by recasting the component equations in other directions. The most useful directions are the tangential and normal directions, and therefore the tangential and normal components of equation (3.2) are needed.

3.3 Acceleration Components

In rectangular co-ordinates

$$\frac{d^2\mathbf{r}}{dt^2} = \frac{d^2x}{dt^2}\hat{\mathbf{i}} + \frac{d^2y}{dt^2}\hat{\mathbf{j}}$$

However, for the normal and tangential components it is necessary to introduce unit vectors $\hat{\boldsymbol{\tau}}$ and $\hat{\mathbf{n}}$. Suppose that ψ is the angle made by $\hat{\boldsymbol{\tau}}$ with $\hat{\mathbf{i}}$, then $\hat{\boldsymbol{\tau}} = \hat{\mathbf{v}}$ (see Figure 3.2). Since

$$\mathbf{v} = v\hat{\boldsymbol{\tau}}$$

differentiation with respect to time yields

$$\frac{d^2\mathbf{r}}{dt^2} = \frac{dv}{dt}\hat{\boldsymbol{\tau}} + v\frac{d\hat{\boldsymbol{\tau}}}{dt}$$

(Note that $\hat{\boldsymbol{\tau}}$ changes with time, not in magnitude of course but in direction). It is clear from Figure 3.2 that

$$\hat{\boldsymbol{\tau}} = \cos\psi\,\hat{\mathbf{i}} + \sin\psi\,\hat{\mathbf{j}}$$

$$\hat{\mathbf{n}} = -\sin\psi\,\hat{\mathbf{i}} + \cos\psi\,\hat{\mathbf{j}}$$

in which $\hat{\mathbf{i}}$ and $\hat{\mathbf{j}}$ are constant. Therefore

$$\frac{d\hat{\boldsymbol{\tau}}}{dt} = -\frac{d\psi}{dt}\sin\psi\,\hat{\mathbf{i}} + \frac{d\psi}{dt}\cos\psi\,\hat{\mathbf{j}}$$

$$= \frac{d\psi}{dt}\hat{\mathbf{n}}$$

and so

$$\frac{d^2\mathbf{r}}{dt^2} = \frac{dv}{dt}\hat{\boldsymbol{\tau}} + v\frac{d\psi}{dt}\hat{\mathbf{n}} \tag{3.4}$$

The magnitude of the tangential component is $dv/dt = d^2s/dt^2$, since the speed $v = ds/dt$ where s is the arc-length. The magnitude of the normal component is

$$v\frac{d\psi}{dt} = v\frac{d\psi}{ds}\frac{ds}{dt} = v^2\frac{d\psi}{ds} = \frac{v^2}{\rho}$$

where $\rho = ds/d\psi$ is the radius of curvature of the trajectory.

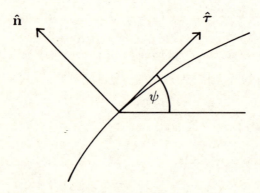

Figure 3.2 The unit vectors $\hat{\boldsymbol{\tau}}, \hat{\mathbf{n}}$

3.4 Power-law Solution

Equation (3.4) can be used to decompose equation (3.2) into

$$\frac{dv}{dt} = -g\sin\psi - kv^n$$

and

$$v\frac{d\psi}{dt} = -g\cos\psi$$

The usefulness of these forms compared with equations (3.3) is evident, since one of them is independent of the drag force. If the above equations are non-dimensionalised using equations (1.12) then they can be written as

$$\frac{dV}{dT} = -\sin\psi - \epsilon V^n \tag{3.5}$$

$$V\frac{d\psi}{dT} = -\cos\psi \tag{3.6}$$

where $\epsilon = kv_0^n/g$. In this general power-law context ϵ still can be interpreted as the ratio of the initial drag to the weight of the projectile. To obtain the position of the projectile at any time equations (3.5) and (3.6) are solved plus $dX/dT = V\cos\psi$ and $dY/dT = V\sin\psi$ subject to initial conditions $V = 1$, $\psi = \alpha$, $X = 0$, $Y = 0$ when $T = 0$. This can only be partially accomplished as will now be demonstrated.

Suppose that equation (3.5) is divided by equation (3.6) then

$$\frac{1}{V}\frac{dV}{d\psi} = \tan\psi + \epsilon V^n \sec\psi$$

which can be rewritten as

$$\frac{dV}{d\psi} - V\tan\psi = \epsilon V^{n+1}\sec\psi$$

This equation is a Bernoulli-type first-order differential equation. It was this very projectile problem which led Jacob Bernoulli (1654 - 1705) to formulate the equation and Gottfried Leibniz (1646 - 1716) to develop the following solution technique in 1696. Division by V^{n+1} yields

$$\frac{1}{V^{n+1}}\frac{dV}{d\psi} - \frac{\tan\psi}{V^n} = \epsilon\sec\psi$$

The change of dependent variable $W = V^{-n}$ produces $dW/d\psi = -nV^{-n-1}\,dV/d\psi$, and so the equation becomes

$$\frac{dW}{d\psi} + (n\tan\psi)W = -n\epsilon\sec\psi$$

This is a linear first-order differential equation with integrating factor $|\sec^n\psi|$, but the physics and geometry of the problem ensures that $-\pi/2 \le \psi \le \pi/2$ and so $\sec^n\psi$ may be used as the integrating factor. Thus

$$\frac{d}{d\psi}(W\sec^n\psi) = -n\epsilon\sec^{n+1}\psi$$

Integration with respect to ψ from α to a general ψ yields

$$\frac{\sec^n\psi}{V^n} - \sec^n\alpha = -n\epsilon\int_\alpha^\psi \sec^{n+1}\xi\,d\xi$$

and so

$$V^n = \frac{\sec^n\psi}{\sec^n\alpha - n\epsilon\int_\alpha^\psi \sec^{n+1}\xi\,d\xi} \qquad (3.7)$$

If v and ψ are determined experimentally (for example, on the ballistics range) equation (3.7) can be used to predict n or ϵ once one of them is known.

When n is an integer the integral in the denominator of equation (3.7) can be expressed analytically in terms of logarithmic and trigonometric functions. Otherwise the integral has to be evaluated by numerical methods such as Simpson's rule, for any given value of ψ (see Kreyszig p.789).

Therefore a partial solution to equations (3.5) and (3.6) has now been given in the form $V = V(\psi)$. Further details about the trajectory would be revealed if the functional forms $\psi = \psi(T)$, $X = X(T)$ and $Y = Y(T)$ could also be found.

From equation (3.6)

$$\frac{dT}{d\psi} = -V\sec\psi$$

Since $\psi = \alpha$ and $V = 1$ when $T = 0$ it is seen by integration that

$$T = -\int_\alpha^\psi V\sec\xi\,d\xi \qquad (3.8)$$

and so T is known as a function of ψ, which is enough for the purposes required. In particular, if $\psi = \psi_f$ is the angle of the projectile when it reaches the horizontal plane through the point of projection,

$$T_f = \int_{\psi_f}^{\alpha} V \sec \xi \, d\xi \tag{3.9}$$

gives the dimensionless time of flight, using equation (3.7). It will be revealed how to obtain ψ_f shortly.

Now $\quad dX/dT = V \cos \psi \quad$ and $\quad dT/d\psi = -V \sec \psi; \quad$ hence

$$\frac{dX}{d\psi} = \frac{dX}{dT} \frac{dT}{d\psi}$$

$$= -V^2$$

Therefore

$$X = -\int_{\alpha}^{\psi} V^2 d\xi \tag{3.10}$$

and the dimensionless range X_f is given by

$$X_f = \int_{\psi_f}^{\alpha} V^2 d\xi. \tag{3.11}$$

Similarly $dY/dT = V \sin \psi$ leading to

$$Y = -\int_{\alpha}^{\psi} V^2 \tan \xi \, d\xi \tag{3.12}$$

In particular ψ_f occurs when $Y = 0$, so the integral equation that determines ψ_f is

$$0 = \int_{\psi_f}^{\alpha} V^2 \tan \xi \, d\xi \tag{3.13}$$

With this solution as ψ_f in equations (3.9) and (3.11) the time of flight and range on the horizontal plane can be determined by quadratures. The integral expressions (3.10) and (3.12) for X and Y can be used to plot any number of points on the trajectory for motion in a medium with a power-law resistance.

Example 3.1

For projectile motion under an $(n = 2)$ power law of resistance show that the relation between V and ψ is

$$\frac{1}{V^2 \cos^2 \psi} = -\epsilon \left[\ln\left(\frac{1 + \sin\psi}{\cos\psi}\right) + \frac{\sin\psi}{\cos^2\psi} + \text{constant} \right]$$

Hence find the equation of the trajectory as a relation between arc-length s and ψ.

Solution

Since ψ lies between $\pm\pi/2$ then for $n = 2$ the power-law formulae yield

$$\int_\alpha^\psi \sec^3 \xi\, d\xi = \frac{1}{2}\{\sec\psi \tan\psi + \ln(\sec\psi + \tan\psi) - \sec\alpha \tan\alpha - \ln(\sec\alpha + \tan\alpha)\}$$

which shall be written in the form $\frac{1}{2}\{f(\psi) - f(\alpha)\}$. Thus equation (3.7) becomes

$$V^2 = \frac{\sec^2\psi}{\sec^2\alpha + \epsilon\{f(\alpha) - f(\psi)\}}$$

which on rearranging becomes

$$\frac{1}{V^2 \cos^2 \psi} = -\epsilon\left[\ln(\sec\psi + \tan\psi) + \sec\psi \tan\psi + \text{constant}\right]$$

$$= -\epsilon\left[\ln\left(\frac{1+\sin\psi}{\cos\psi}\right) + \frac{\sin\psi}{\cos^2\psi} + \text{constant}\right]$$

as required. Now $V = dS/dT$ where S is the non-dimensional arc-length with $s = v_0^2 S/g$. Therefore

$$\frac{dS}{d\psi} = V\frac{dT}{d\psi}$$

$$= -V^2 \sec\psi$$

$$= \frac{-\sec^3\psi}{\sec^2\alpha + \epsilon\{f(\alpha) - f(\psi)\}}$$

Integration with respect to ψ yields

$$S = \frac{1}{2\epsilon}\ln\left[\sec^2\alpha + \epsilon\{f(\alpha) - f(\psi)\}\right] + \frac{1}{2\epsilon}\ln\sec^2\alpha$$

if $S = 0$ when $\psi = \alpha$. In dimensional form this is

$$s = \frac{1}{2k}\ln\left[1 + \frac{kv_0^2\cos^2\alpha}{g}\{f(\alpha) - f(\psi)\}\right]$$

Example 3.2

A projectile is fired at an angle of $60°$ to the horizontal with an initial speed of 40 ms^{-1} in a medium whose drag per unit mass is $0.01v^2$. Determine the range and time of flight.

Solution

When $n = 2$ the dimensional form of the solution (3.7) becomes

$$v^2 = \frac{\sec^2\psi}{\sec^2\alpha/v_0^2 + \{f(\alpha) - f(\psi)\}\,k/g}$$

using the previous example. Now $v_0 = 40$, $\alpha = 60°$ and $k = 0.01$, and so

$$V = \frac{0.7825\sec\psi}{\sqrt{7.23 - \sec\psi\tan\psi - \ln(\sec\psi + \tan\psi)}}$$

Using this and equation (3.13) it is found by a trial and error procedure using Simpson's rule that $\psi_f = -1.26$(radians). Then equations (3.9) and (1.12) yield

$$t_f = -\frac{1}{g}\int_{\pi/3}^{-1.26} v\,\sec\xi\,d\xi \;=\; 5.5\;\text{(seconds)}$$

while equation (3.11) produces

$$x_f = \frac{1}{g}\int_{-1.26}^{\pi/3} v^2 d\xi \;=\; 64.9\;\text{(metres)}$$

Note that for motion in a drag-free medium the corresponding results are a time of flight of 8.2 seconds and a range of 141.4 metres, illustrating that a power-law resistance also reduces the no-drag predictions.

Example 3.3

Equipment is to be dropped by parachute from an aircraft flying horizontally with speed v_0. The best time for the parachute to open is when the equipment's speed is a minimum. Before the parachute opens the air offers a resistance kv^2 per unit mass, where k is constant. Obtain an expression for this best time in terms of v_0, k, and g.

Solution

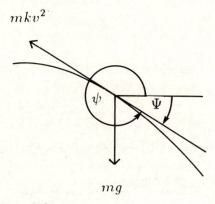

mkv^2

mg

Figure 3.3

Before the parachute opens the equipment's motion is governed by the following equation in dimensional form:

$$m\frac{dv}{dt} = mg\sin\Psi - mkv^2 \quad \text{(tangential direction)}$$

$$mv\frac{d\Psi}{dt} = mg\cos\Psi \quad\quad \text{(normal direction)}$$

where Ψ is the angle below the horizontal as in Figure 3.3. Elimination of t yields

$$\frac{dv}{d\Psi} = v\tan\Psi - \frac{kv^3}{g}\sec\Psi$$

The solution of this differential equation is

$$\frac{\sec^2\Psi}{v^2} = \frac{k}{g}\left\{\ln\left[\frac{1+\sin\Psi}{\cos\Psi}\right] + \frac{\sin\Psi}{\cos^2\Psi}\right\} + \frac{1}{v_0^2} \tag{3.14}$$

since $v = v_0$ when $\Psi = 0$. Now v will be a maximum or a minimum when $dv/dt = 0$. This gives

$$v^2 = \frac{g}{k} \sin \Psi$$

directly from the equation for the tangential direction. Also

$$\frac{d^2v}{dt^2} = g \cos \Psi \frac{d\Psi}{dt} - 2kv \frac{dv}{dt}$$

$$= \frac{g^2 \cos^2 \Psi}{v}$$

when $dv/dt = 0$. Thus $d^2v/dt^2 > 0$, and therefore v has a minimum value when $v = \sqrt{g \sin \Psi / k}$. Substitution into equation (3.14) yields Ψ_m, the angle for minimum v as the solution of the transcendental equation

$$\ln(\sec \Psi + \tan \Psi) = \operatorname{cosec} \Psi - \frac{g}{kv_0^2}$$

The normal direction equation, using t_0 as the time for opening produces

$$t_0 = \frac{1}{\sqrt{kg}} \int_0^{\Psi_m} \frac{\sec^2 \Psi d\Psi}{\sqrt{\ln(\sec \Psi + \tan \Psi) + \sec \Psi \tan \Psi + g/(kv_0^2)}}$$

3.5 Applications of the Power-law Solution

(i) **The Short-Arc Method**   Early work on trajectories of cannon balls assumed that the drag was proportional to the square of the speed. With the introduction of rifling and the use of elongated projectiles it became necessary to make an experimental study of the laws of resistance of these new projectiles. It was soon evident that the square of the speed or any other single power could not be used. Mayevski (1872) suggested however that it was possible to express the retardation as proportional to a power of the speed in a restricted velocity zone. A good agreement with experimental firings was obtained when the drag was written in the form $A_n v^n$, where n and A_n are different integer constants for different velocity ranges or, equivalently, different short arcs of the trajectory. For example, in gunnery, $n = 2$ from the lowest muzzle velocity up to 240 ms^{-1}, $n = 3$ from 240 to 295 ms^{-1}, $n = 5$ from 295 to 375 ms^{-1}, with different resistance coefficients in each region.

In each region the solution is obtained by the Bernoulli method for drag proportional to v^n. Since the speed of the projectile decreases during its flight it may pass from a region with one n-value to a region with a lower n-value. When the lower-bound speed for a particular n-value is reached, the angle of the projectile is evaluated. This and the lower-bound speed become the initial conditions for the new region. Thus the Bernoulli solutions are patched together at the junction points, and if the tangential accelerations are equated a new A_n is determined from

$$A_n v^n = A_{n_0} v^{n_0}$$

where n_0 is the previous value of n. The value for n_0 is frequently $(n+1)$, but it does not necessarily have to be.

Since the Bernoulli results are always expressed in terms of quadratures a new set of quadratures is required for each short arc. The numerical computation is considerable but the short-arc method is capable of producing correct results for many different types of trajectories including those with high angles and high muzzle speeds.

(ii) **Euler's Method**   Frequently there are obstacles between a gun and the enemy, so weapons have been developed which enable the projectile to go over the obstacle and deliver a plunging-fire on the enemy. These weapons are usually characterized by a low initial velocity and a high angle of departure, as for example with trench mortars. Since the velocity of the projectile is low its trajectory will not be very long and therefore cannot extend through a great depth of the atmosphere. Thus it is possible to assume constant atmospheric density and, as long the projectile's velocity is less than 240 ms^{-1}, it is a reasonable approximation to take the drag to be proportional everywhere to the square of the speed. These approximations were first suggested by Euler (1707 - 1783). In this case equation (3.7) gives

$$V^2 = \frac{\sec^2 \psi}{\epsilon\{K - f(\psi)\}} \tag{3.15}$$

where

$$f(\psi) = \sec\psi \tan\psi + \ln\left(\sec\psi + \tan\psi\right)$$

and

$$K = f(\alpha) + \frac{g \sec^2 \alpha}{kv_0^2}$$

Substitution into equations (3.8), (3.10) and (3.12) and conversion to dimensional form yields

$$\sqrt{kg}\, t = \int_\psi^\alpha \sec^2 \xi \{K - f(\xi)\}^{-\frac{1}{2}} d\xi$$

$$kx = \int_\psi^\alpha \sec^2 \xi \{K - f(\xi)\}^{-1} d\xi$$

$$ky = \int_\psi^\alpha \sec^2 \xi \tan \xi \{K - f(\xi)\}^{-1} d\xi$$

Tables of the above integrals were prepared by J.C.F. Otto and used extensively before World War I.

Equation (3.6) suggests that as t increases, ψ decreases. Now consider the right-hand side of equation (3.15) in the limit as ψ tends to $-\pi/2$. Using L'Hôpital's rule

$$\lim_{\psi \to -\pi/2} \left[\frac{\sec^2 \psi}{\epsilon\{K - f(\psi)\}} \right] = \lim_{\psi \to -\pi/2} \left[\frac{2 \sec^2 \psi \tan \psi}{\epsilon(-2 \sec^3 \psi)} \right]$$

$$= \lim_{\psi \to -\pi/2} \left[\frac{-\sin \psi}{\epsilon} \right]$$

$$= \frac{g}{kv_0^2}$$

Therefore

$$\lim_{\psi \to -\pi/2} v^2 = \frac{g}{k}$$

and so v is finite as $\psi \to -\pi/2$ with a value equal to the terminal speed.

Now the integral in equation (3.10) is finite, but the integrals contained in equations (3.8) and (3.12) are both infinite as $\psi \to -\pi/2$. Thus the trajectory has a vertical asymptote in the direction of increasing t, showing that x (and therefore the horizontal range) is strictly bounded (Figure 3.4). This behaviour is very different from the no-drag case, but similar to the case $n = 1$.

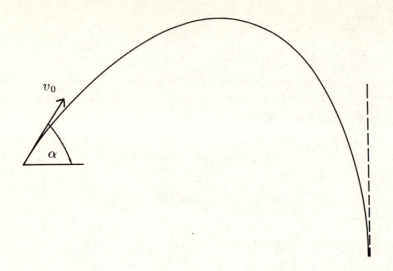

Figure 3.4 The trajectory and asymptote for a mortar projectile

3.6 Other Non-linear Drag Functions

Consider now the more general case where the magnitude of the drag force is not proportional to v^n. Using equation (3.1) it is seen that the equation for the motion of the projectile in dimensional form is

$$\frac{d^2\mathbf{r}}{dt^2} = -g\hat{\mathbf{j}} - \frac{\rho A v^2 C_D}{2m}\hat{\mathbf{v}} \tag{3.16}$$

The mathematics is simplified by decomposing equation (3.16) in the two directions which isolate the forces, namely the $\hat{\mathbf{n}}$ and $\hat{\mathbf{i}}$ directions. This produces equation (3.6) in dimensional form

$$v\frac{d\psi}{dt} = -g\cos\psi$$

and the horizontal-component equation

$$\frac{d^2x}{dt^2} = -\frac{\rho A v^2 C_D}{2m}\cos\psi \tag{3.17}$$

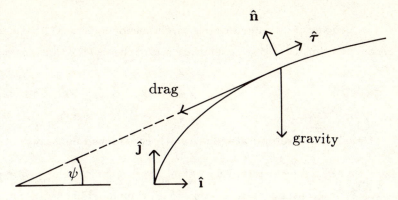

Figure 3.5 Drag and gravity forces in relation to unit vectors

Now

$$\frac{d}{d\psi}\left[\frac{dx}{dt}\right] = \frac{d^2x}{dt^2}\Big/\frac{d\psi}{dt}$$

$$= \frac{\rho A v^3 C_D}{2mg}$$

But $dx/dt = v\cos\psi$, leading as before {see equation (3.10)} to

$$\frac{dx}{d\psi} = -\frac{v^2}{g}$$

and therefore

$$\frac{dy}{d\psi} = -\frac{v^2}{g}\tan\psi$$

Hence equation (3.16) can be investigated by considering the equations

$$\frac{d(v\cos\psi)}{d\psi} = \frac{\rho A v^3 C_D}{2mg} \tag{3.18}$$

and

$$\frac{dy}{d\psi} = -\frac{v^2}{g}\tan\psi \tag{3.19}$$

where ρ is a function of y and C_D is a function of both v and y. When these are solved for v and y as functions of ψ the remaining information about the trajectory can be obtained by quadratures using the dimensional forms of equations (3.8) and (3.10).

A survey of the various classes of solution for equations (3.18) and (3.19) was presented by Leimanis (1958). A summary is given here of the three cases considered.

Case 1   Density and temperature are both considered constant

When the temperature is constant C_D is assumed to only vary with v. If the density is also constant, then equation (3.18) is replaced by

$$\frac{d(v \cos \psi)}{d\psi} = f_1(v)$$

the so-called equation of the hodograph. The problem of determining the forms of $f_1(v)$ for which the equation is integrable was completely solved by Drach (1920). Only a few of the forms he obtained can be considered as approximations to actual circumstances. The others do not represent actual drag functions and are therefore only of theoretical interest. One example of a useful form is kv^{n+1}/g treated earlier in this chapter in detail.

Case 2   Density variable but temperature kept constant

Equation (3.18) then has the form,

$$\frac{d(v \cos \psi)}{d\psi} = \rho(y)f_2(v)$$

and the integrable functions without their multiplying constants are shown in Table 3.1.

$f_2(v)$	$\rho(y)$
v	$\exp(-cy), c$ constant
v	$(1 + cy)^{-\frac{1}{2}}$
v	$1 + cy$
v	$(1 + cy)^{-2}$
v^2	$(1 + cy)^{-1}$
$v^{-2}\exp(-cv^2/(2g))$	$\exp(-cy)$

Table 3.1 Forms of f_2 and ρ for integrability of equation (3.16)

Case 3   Density and temperature both variable

Suppose that

$$\rho(y) = \rho_0 \left(1 - \frac{\lambda y}{\tau_0} \right)^{(1+R\lambda)/(R\lambda)}$$

$$\tau(y) = \tau_0 - \lambda y$$

where $\lambda \approx 0.005$, R is the gas constant, $\tau(y)$ is the absolute temperature of the air at height y, and τ_0 is the standard value of the absolute air temperature at sea-level $(y = 0)$. These laws correspond to actual conditions in nature and, using them, equation (3.18) becomes

$$\frac{d(v\cos\psi)}{d\psi} = \left(1 - \frac{\lambda y}{\tau_0} \right)^{(2+R\lambda)/(2R\lambda)} f_3(w)$$

where $w = v\left(1 - \lambda y/\tau_0\right)^{-\frac{1}{2}}$. The integrable functions are

$$f_3(w) = w$$

$$f_3(w) = \frac{1}{w^2}\left\{ 1 - \frac{\lambda w^2}{2g\tau_0} \right\}^{(1+R\lambda)/(R\lambda)}$$

$$f_3(w) = w\left\{ 1 - \frac{\lambda w^2}{2g\tau_0} \right\}^{\frac{1}{2}}$$

3.7 Exercises

1. Write down the equations of motion in the normal and tangential directions for a projectile travelling in a resistance-free medium (that is, gravity only). If $\psi = \alpha$, $v = v_0$ when $t = 0$ solve the equations to obtain expressions for v, t, x and y as functions of ψ at any point on the trajectory. Hence find relations between

    (i) x and t (ii) y and t (iii) y and x.

Attempt the analysis using dimensional variables, and compare with a similar analysis using dimensionless variables.

2. For a projectile travelling in a medium whose resistance is proportional to the speed use the approach in this chapter to find the relation between V and ψ under the usual initial conditions. Hence find a relation between T and ψ, and show that

$$\frac{dX}{dT} = e^{-\epsilon T} \cos \alpha$$

$$\frac{dY}{dT} = e^{-\epsilon T} \sin \alpha + \frac{(e^{-\epsilon T} - 1)}{\epsilon}$$

3. A bombing aeroplane flies at a speed of 1080 km h^{-1} at a height of 10 km. At what distance from the target should the bombardier release the bomb, assuming that the drag force on the bomb is proportional to the square of its speed with a resistance coefficient per unit mass of 10^{-5} m^{-1}?

4. A projectile is fired with initial speed 400 ms^{-1} at an angle of $\pi/4$ radians in a medium whose resistance per unit mass is $0.0002v^4$. Determine the time of flight and the range for impact on the horizontal plane through the projection point. What is the striking velocity?

5. For trench mortars the drag is proportional to the square of the speed. Using dx/dt and ψ as the dependent variables write down the normal and horizontal equations of motion for a trench mortar shell. Solve them to show that

$$\left[\frac{dt}{dx}\right]^2 = \frac{1}{v_0^2 \cos^2 \alpha} - \frac{k}{g}\{f(\psi) - f(\alpha)\}$$

in the usual notation. Hence show that

$$x = \frac{1}{k} \int_{\psi}^{\alpha} \sec^2 \xi \, \{K - f(\xi)\}^{-1} \, d\xi.$$

6. For a projectile under an $(n = 3)$ power law of resistance obtain expressions for the time of flight and the range.

7. Describe how to determine the range on an inclined plane of angle β for a projectile under a v^n power law of resistance.

8. Use the Mayevski criteria for drag with the initial $k = 0.001$ to calculate the time of flight and the range for a projectile fired in air with initial speed 280 ms^{-1} at an angle $\pi/3$ radians to the horizontal and impacting on the horizontal plane through the projection point.

9. For an $(n = 2)$ power law of resistance calculate the curve of safety.

10. The breaking of the Möhne and Eder Dams by 617 Squadron during World War II was achieved through the idea of skipping a bomb over the water surface and measuring where it came to rest. Investigate this for an aeroplane flying at a speed 386 km h^{-1} and releasing the bomb horizontally at an altitude 18 metres above the water surface. Assume three skips of the bomb and a coefficient of restitution $e = 0.5$ between the bomb and the water surface. After the third skip the bomb sinks through the water and should come to rest at a depth 42 metres below the surface and up against the dam wall. Assume that the air offers a resistance $0.0003 \, v^2$ per unit mass while the water offers a resistance $0.23 \, v^2$ per unit mass. How far from the wall should the bombardier release the bomb?

11. Multiply equation (3.6) by $\cos \psi$. With $n = 2$ in equation (3.5) make the substitution $\zeta = \sin \psi$ in equation (3.5) and the new equation (3.6). Solve the resulting equations to obtain $v = v(\zeta)$.

12. Choose one of the combinations for f_2 and ρ in Table 3.1 and solve equations (3.18) and (3.19).

13. Show that equation (3.18) is integrable for each of the functions given in Case 3.

4. THE BASIC EQUATIONS AND THEIR NUMERICAL SOLUTION

"I shot an arrow into the air,

It fell to earth, I know not where."

H.W. Longfellow (1807 - 1882)

4.1 The Basic Equations

The air resistance acting on a projectile can be separated into four categories - forebody drag, base drag, skin friction and protuberance drag. Forebody drag arises because some of the energy of the projectile is used to compress the air in front of the projectile and in some cases to form shock waves. Base drag is due to the turbulent wake behind the projectile and is more pronounced if the rear of the projectile is blunt. Skin friction is caused by air adhering to the surface of the projectile. It can be reduced by polishing the surface. The protuberance drag is really a combination of the forebody drag, base drag and skin friction on any protuberance attached to the main projectile body. For example the principal protuberance on a shell fired from a gun is the driving band.

When the projectile's speed (v) is divided by the local speed of sound in air (a) this nondimensional ratio is called the Mach number (M). Near $M = 1$ (the transonic region) the forebody drag increases dramatically, and therefore becomes effectively the total aerodynamic drag. Since a is proportional to the square root of the absolute temperature of the air, and the drag coefficient C_D introduced at the beginning of Chapter 3 also depends on air temperature, it follows that C_D is a function of the Mach number for a given projectile (see Figure 4.1). In general C_D has to be determined experimentally for each type of projectile under consideration.

For many projectiles travelling at speeds below $M = 0.6$ the value of C_D can be taken as constant, and this is also true for streamlined shells travelling above $M = 6$. In these cases the results of Chapter 3 with drag proportional to v^2 can be applied when ρA is also constant.

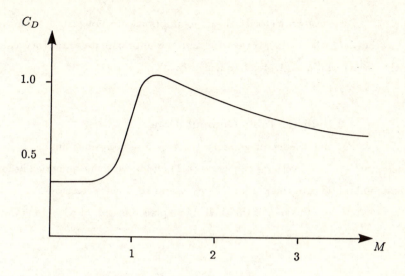

Figure 4.1 Variation of C_D with Mach number for a shell projectile

However for a non-spinning projectile travelling at any speed, the basic equations for the motion of its centre of gravity are written as

$$m\frac{d^2\mathbf{r}}{dt^2} = m\mathbf{g} - \frac{1}{2}\rho(y)Av^2C_D(M)\hat{\mathbf{v}} \tag{4.1}$$

with $\mathbf{v} = \mathbf{v}_0$, $\mathbf{r} = \mathbf{0}$ when $t = 0$. This problem is well-posed, and it has been shown that it has a unique solution for each set of problem parameters. This solution gives a very good approximation to the actual motion of the projectile.

In effect these basic equations governing the approximate trajectory have been obtained by making the following simplifying assumptions:

(i) The projectile moves with its cross-sectional area normal to the trajectory tangent.

(ii) The only forces acting are constant gravity and variable drag.

(iii) The density and temperature are functions only of the height above the Earth's surface.

(iv) There is no wind.

(v) The rotation of the Earth is neglected.

For most trajectories these assumptions are not strictly true, but the difference between the actual trajectory and the approximate trajectory calculated using the basic equations (4.1) is frequently small.

4.2 Ballistic Table Computations

For the science of gunnery the above assumptions enable ballisticians to concentrate on the governing equations (4.1) and solve them numerically for many different combinations of the parameters v_0, α and shell characteristics.

Usually the area of the shell A is replaced by $\pi d^2/4$ where d is the calibre (diameter), and the equation (4.1) is rewritten as

$$\frac{d^2 \mathbf{r}}{dt^2} = \mathbf{g} - \rho(y)\frac{d^2}{m}v^2 K_D(M)\hat{\mathbf{v}} \qquad (4.2)$$

where $K_D = \pi C_D/8$ is called the Siacci function. With the appropriate initial conditions the equations (4.2) are converted to a system of first-order equations and integrated numerically using a multi-step iterative procedure (Runge-Kutta). In general the values of the function $K_D(M)$ are estimated from tables of a G-function where

$$G\left(\frac{v}{a}\right) = \rho_0 \frac{v}{a} K_D\left(\frac{v}{a}\right)$$

It will shortly be convenient (see equation (4.3)) to introduce a function $E(y, v)$ defined by

$$E(y, v) = \rho_0 exp(-0.0001036y)K_D(M)\frac{vd^2}{m}$$

Now the air temperature at various heights is taken from Table 4.1 which enables $a(y) = (\gamma R\tau)^{\frac{1}{2}}$ to be calculated, where γ is the ratio of specific heats, R is the gas-constant and τ is the absolute temperature. Thus E takes the form

$$E(y, v) = exp(-0.0001036y)G\left(\frac{v}{a}\right)\frac{ad^2}{m}$$

Tables of the G function are given in McShane, Kelley and Reno (1953) pp. 810-818.

If the step-length is small enough and the assumptions (i) - (v) are valid, the difference between the actual trajectory and the approximate trajectory is quite small, and can be estimated by a perturbation theory which considers small corrections for the Coriolis effect, gravity variations, wind, drift, lift and variable density (see Chapter 6). In gunnery these corrections are included in range tables.

With the above assumptions the projectile remains in the vertical plane containing the initial velocity vector, and so the problem is still 2-dimensional. With regard to assumption (iii), standard relations are often assumed for the density and the speed of sound. Standard density values $\rho(y)$ are given by international agreement such that on any given day the actual density is unlikely to differ very much from the standard value.

For temperate latitudes Batchelor (1972) gives the following table.

Height above sea-level (m)	Density (kg m^{-3})	Temperature (°C)
0	1.226	15.0
500	1.168	11.7
1,000	1.112	8.5
1,500	1.059	5.2
2,000	1.007	2.0
3,000	0.910	-4.5
4,000	0.820	-11.0
5,000	0.736	-17.5
6,000	0.660	-24.0
8,000	0.525	-37.0
10,000	0.413	-50.0
12,000	0.311	-56.5
14,000	0.227	-56.5
16,000	0.165	-56.5
18,000	0.121	-56.5

Table 4.1 Average values of density and temperature for the Standard Atmosphere in temperate latitudes

Some ballisticians replace the standard density table by an analytic expression of the form

$$\rho(y) = \rho_0 exp(-0.0001036y)$$

where y must be in metres and ρ_0 is the standard density at sea-level.

With these standard density and temperature functions plus relevant experimental values of the drag coefficient C_D (see Figure 4.1) it is possible to obtain highly precise solutions of equations (4.2) using a Runge-Kutta technique. The IMSL package DVERK is useful for undergraduates but more sophisticated programmes are used by ballisticians (see for example Morrey's method as modified by McShane, Kelley and Reno(1953)).

Essentially equation (4.2) can be rewritten in its Cartesian components as

$$\left. \begin{array}{l} \dfrac{d^2x}{dt^2} = -E\dfrac{dx}{dt} \\[2ex] \dfrac{d^2y}{dt^2} = -E\dfrac{dy}{dt} - g \end{array} \right\} \qquad (4.3)$$

where for gunnery problems

$$E(y,v) = \rho_0 exp(-0.0001036y)K_D(M)\frac{vd^2}{m}$$

In applying the Runge-Kutta technique these two coupled second-order differential equations are converted to a system of four first-order differential equations by writing $dx/dt = v_x, dy/dt = v_y$, leading to

$$\frac{dv_x}{dt} = F_1(v_x, v_y, y)$$

$$\frac{dv_y}{dt} = F_2(v_x, v_y, y)$$

One iterative application of the computer package for the initial value problem governed by these two equations with $v_x(0) = v_0 \cos\alpha, v_y(0) = v_0 \sin\alpha$ produces numerical values of v_x and v_y, at each time-step interval. A second application produces corresponding x and y values. In general there appears to be little advantage gained numerically by converting equations (4.3) to non-dimensional form.

As in all numerical procedures of this type considerable experience and skill are needed to determine the length of the time-step interval which will produce the accuracy required without exceeding the capabilities of the computer.

The ballistic table for a certain shell can thus be built up by repeating the process for many different variations of the parameters v_0, α and d^2/m.

4.3 Simple Application

An illustration of the usefulness of ballistic tables is obtained by considering the ballistic table for range (x_f) and time of flight (t_f) for a non-spinning spherical ball moving through air (Table 4.2).

b_1	b_2	b_3
0	0	0
2	0.18	0.20
4	0.34	0.40
6	0.48	0.59
8	0.61	0.78
10	0.73	0.96
12	0.84	1.13
14	0.95	1.30
16	1.05	1.47
18	1.14	1.63
20	1.22	1.79
22	1.30	1.95
24	1.38	2.11
26	1.45	2.26
28	1.52	2.41
30	1.58	2.56

Table 4.2 Ballistic table for a non-spinning spherical projectile

The headings at the top of each column are defined by

$$b_1 c = v_0^2 \sin 2\alpha$$

$$b_2 c = x_f$$

$$b_3 c = v_0 t_f \cos \alpha$$

where the ballistic coefficient for the sphere is given by $c = m/(C_D d^2)$ which is not dimensionless. Intermediate values of b_1, b_2 and b_3 can be obtained by linear interpolation within the table. A more complete table is given in Daish (1972) using different notation.

When using this ballistic table the procedure to determine the range x_f (metres) and the time of flight t_f (seconds) is very straight-forward. From knowledge of the drag co-efficient C_D for a particular sphere plus its mass and diameter a value of the ballistic coefficient c is calculated. If v_0 and α are known then b_1 can be calculated also. The ballistic table can then be used to read off b_2 and b_3, and so x_f and t_f can be obtained. The following two examples illustrate this.

Example 4.1

What is the range and time of flight of a soccer ball kicked off the ground at an angle of $50°$ to the horizontal with an initial speed 28 ms^{-1}?

Solution

The trajectory of the ball is only considered until its first bounce occurs. The mass of a soccer ball is 0.42 kg and its diameter is 0.22 m. Since its diameter is large it is easy to calculate that its Reynolds number at this initial speed is well above the critical level at which turbulence occurs (see Chapters 7 and 8) and hence $C_D = 0.2$. Thus the ballistic coefficient is $c = 43$.

With $v_0 = 28$, $\alpha = 50°$ it is seen that $b_1 = 17.1$. By interpolation in the ballistic tables $b_2 = 1.10$ and $b_3 = 1.56$.

Hence the range is approximately 47.30 m and the corresponding time of flight is 3.72 s.

Example 4.2

A baseball pitching machine projects baseballs at a speed of 35 ms^{-1}. At what angle should it be set to give catches to the outfielders at a distance 70 m from the machine?

Solution

The mass of a baseball is 0.15 kg and its diameter is 0.07 m. The size of a baseball is such that at this speed the critical level has not been reached and flow is laminar (again see Chapters 7 and 8). This means that the drag coefficient C_D will be taken as 0.45. Therefore the ballistic coefficient is $c = 68.03$.

Since $x_f = 70$ a simple calculation yields $b_2 = 1.03$.

From the ballistic table $b_1 = 15.6$ and $b_3 = 1.43$. Therefore

$$\sin 2\alpha = \frac{b_1 c}{v_0^2} = 0.866$$

yielding $\alpha = 30.0°$. The time of flight is then 3.21 s.

4.4 Range Tables

The data required for proper aiming of weapons are contained in range tables. These refer not only to the weapon but also to the particular kind of ammunition and the kind of air structure at the time of firing. Thus a range table (firing or bombing) is made for the person using the weapon, and its basic purpose is to produce a hit on the target with that weapon. The ballistic table on the other hand is made for the use of the person who prepares the range table.

The method of preparing range tables from a ballistic table is restricted principally by the lack of precise knowledge of the drag coefficient of the new projectile for which range tables are required. Firstly a small batch of the new projectiles are sampled and an average Siacci function K_D is determined in the wind tunnel. As this average is sometimes called the Gâvre drag function in gunnery application, it is frequently denoted by K_G.

The ballistic equations (4.3) are then solved numerically with this K_G for many variations of the parameters v_0, α and d^2/m to produce a ballistic table.

Using a pre-determined powder charge the projectile is fired with a fixed initial speed for various angles of elevation α_i. The corresponding ranges x_i are measured. If the ballistic table is consulted for each set of (α_i, x_i) a corresponding drag function K_i is obtained. Although the K_i will most likely not be constant they are usually found to vary only slowly from one set of initial conditions to the next. Hence the drag function can be obtained as a function of the range x by interpolating a smooth curve $K_D(x)$ through the various K_i. Therefore to find the elevation α at which to fire the gun in order to obtain a certain range x it is necessary to obtain $K_D(x)$ from the smooth curve and then use the ballistic table to find α.

Sometimes K_i varies so rapidly that the validity of the whole procedure should be questioned. This is particularly likely with high anti-aircraft shells which almost reach the vertex before exploding, and also with some bomb trajectories. In these situations it is necessary to modify the equations (4.2) to include other aerodynamic forces, particularly the lift.

4.5 Variations within the Basic Equations

Frequently the numerical computations required for the solution of equation (4.3) are handled more efficiently by a rearrangement of the system of differential equations.

To show how these rearrangements can be obtained the equations (4.3) are rewritten as

$$\frac{dv_x}{dt} = -Ev_x$$

$$\frac{dv_y}{dt} = -Ev_y - g$$

For any differentiable function $F(t)$ the chain rule yields

$$\frac{dF}{dp} = \frac{dF}{dt}\frac{dt}{dp}$$

and so the basic equations can be recast in the form

$$\frac{dx}{dp} = v_x \frac{dt}{dp} \tag{4.4}$$

$$\frac{dy}{dp} = v_y \frac{dt}{dp} \tag{4.5}$$

$$\frac{dv_x}{dp} = -Ev_x \frac{dt}{dp} \tag{4.6}$$

$$\frac{dv_y}{dp} = -Ev_y \frac{dt}{dp} - g \frac{dt}{dp} \tag{4.7}$$

If the variable σ is defined by

$$\sigma = \frac{dy}{dx} = \frac{v_y}{v_x} = \tan \psi$$

then

$$\frac{d\sigma}{dp} v_x = \frac{dv_y}{dp} - \sigma \frac{dv_x}{dp}$$

$$= -Ev_y \frac{dt}{dp} - g \frac{dt}{dp} + \sigma Ev_x \frac{dt}{dp}$$

$$= -g \frac{dt}{dp}$$

Thus an alternative to using either equation (4.6) or (4.7) is to use

$$\frac{d\sigma}{dp} = -\frac{g}{v_x} \frac{dt}{dp} \tag{4.8}$$

With different choices for p the original basic equation can be recast into new forms, some of which are more convenient to use numerically for special trajectory types.

For example with $p \equiv x$ the equations (4.4), (4.5), (4.6) and (4.8) become respectively

$$\frac{dt}{dx} = \frac{1}{v_x}, \quad \frac{dy}{dx} = \sigma, \quad \frac{dv_x}{dx} = -E, \quad \frac{d\sigma}{dx} = -\frac{g}{v_x^2}$$

The last three equations can be solved independently as a system, and then t can be obtained from the first equation by quadrature.

On the other hand, with $p \equiv \sigma$ the equations (4.8), (4.4), (4.5) and (4.6) become respectively

$$\frac{dt}{d\sigma} = -\frac{v_x}{g}, \qquad \frac{dx}{d\sigma} = -\frac{v_x^2}{g}, \qquad \frac{dy}{d\sigma} = -\frac{\sigma v_x^2}{g}, \qquad \frac{dv_x}{d\sigma} = \frac{E v_x^2}{g}$$

The last two equations can now be solved independently as a system, and t and x determined by quadratures.

Still further, with $p \equiv y$ on the ascending or descending branch of the trajectory the equations (4.5), (4.4), (4.8) and (4.7) become respectively

$$\frac{dt}{dy} = \frac{1}{v_y}, \qquad \frac{dx}{dy} = \frac{1}{\sigma}, \qquad \frac{d\sigma}{dy} = -\frac{g\sigma}{v_y^2}, \qquad \frac{dv_y}{dy} = -E - \frac{g}{v_y}$$

Again the last two equations can be solved independently as a system with t and x determined by quadratures. This last system of equations has applications in the computation of dive-bombing trajectories.

4.6 Graphical Technique

Although graphs can be prepared from the large amount of computation carried out in solving equations (4.3) a simple graphical technique has recently been developed by de Mestre and Catchpole (1988) to help undergraduates understand what the mathematics is doing.

Essentially the differential equation (4.1) is discretized by replacing $d^2\mathbf{r}/dt^2$ by $\left(\mathbf{v}^{(n+1)} - \mathbf{v}^{(n)}\right)/h$ where h is the time-step and $\mathbf{v}^{(n)}$ denotes the velocity vector at time nh $(n = 0, 1, 2, 3, \dots)$. This means that the differential equation (4.1) can be approximately replaced by

$$\mathbf{v}^{(n+1)} = \lambda v_x^{(n)} \hat{\mathbf{i}} + \left\{ \lambda v_y^{(n)} - gh \right\} \hat{\mathbf{j}}$$

where

$$\lambda = 1 - \frac{h\rho A C_D v^{(n)}}{2m}$$

An iterative numerical procedure can then be applied to determine $\mathbf{v}^{(n+1)}$ at each stage. In practice, however, $\mathbf{v}^{(n+1)}$ can be determined graphically using the following procedure.

If $\mathbf{v}^{(n)}$ is known for some time-step, its magnitude is first of all reduced by the magnitude $h\rho A C_D \left\{v^{(n)}\right\}^2 / (2m)$ of the drag effect. This new vector then has its y-component reduced by the gravity effect gh to produce the resultant vector $\mathbf{v}^{(n+1)}$ (see Figure 4.2).

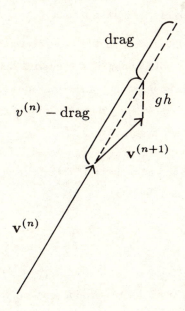

Figure 4.2 Graphical technique for a projectile with drag

A vector-addition diagram consisting of consecutive $\mathbf{v}^{(n)}$ $(n = 0, 1, 2, \dots)$ will represent the approximate path of the trajectory, since the vectors are the tangents to the projectile's path at the beginning of each time step. The magnitude of each vector segment multiplied by the time-step interval h is an approximation to the distance travelled by the projectile during that time-step, because $\mathbf{r}^{(n+1)} - \mathbf{r}^{(n)} = h\mathbf{v}^{(n)}$ in the discretization process. Consequently if $\mathbf{r}^{(0)} = \mathbf{0}$ this difference equation reduces to

$$\mathbf{r}^{(n)} = h \sum_{k=0}^{n-1} \mathbf{v}^{(k)}$$

As in all finite-difference approaches the accuracy of the approximation depends on the length of the time-step h.

One set of criteria for determining the flight characteristics is:

(i) The time to the vertex of the trajectory is the interval (nh) between the start of the trajectory and the beginning of the time-step for which the y-component of $\mathbf{v}^{(n)}$ is first negative.

(ii) The time that the projectile takes to return to the horizontal plane through the initial point is the interval between the start of the trajectory and the beginning of the time-step for which the y-value is first negative (or equivalently when the sum of consecutive y-components of the $\mathbf{v}^{(n)}$ is first negative).

Alternative and more accurate criteria can be based on interpolation of the associated graphical picture.

Example 4.3

With $h = 1$ second, $v_0 = 40$ ms^{-1}, $\alpha = \pi/3$ radians, assume that the projectile is acted on by a drag force proportional to the square of its speed with $\rho A C_D/(2m) = 0.01$ m^{-1}. Determine the maximum height reached, the time of flight and the horizontal range using the approximate de Mestre - Catchpole graphical technique.

Solution

The differential system that is to be solved graphically is

$$\frac{d\mathbf{v}}{dt} = -g\hat{\mathbf{j}} - 0.01v^2\hat{\mathbf{v}}$$

$$\frac{d\mathbf{r}}{dt} = \mathbf{v}$$

First of all the vector $\mathbf{v}^{(0)} = [20, 20\sqrt{3})]$ is drawn illustrating the components of the initial velocity vector.

Its magnitude is then reduced by $0.01 \times (40)^2 \times 1 = 16.0$, which approximates the drag effect during the first one second interval. Finally 9.8×1 is subtracted from the y-component of the reduced vector and so produces $\mathbf{v}^{(1)} = [12.0, 11.0]$, correct to one decimal place. The process is repeated for each time-step (in this case 1 second), and the vector-addition diagram is constructed from consecutive values of $\mathbf{v}^{(n)}$ (see Table 4.3 and Figure 4.3).

n	0	1	2	3	4	5	6	7
$v_x^{(n)}$	20.0	12.0	10.0	9.0	7.8	6.2	4.6	3.3
$v_y^{(n)}$	34.6	11.0	-0.6	-10.3	-18.7	-24.8	-28.3	-30.0
Drag reduction		16.0	2.6	1.0	1.9	4.1	6.5	8.2
$y^{(n)}$	0.0	34.6	45.6	45.0	34.7	16.0	-8.8	-37.1
$x^{(n)}$	0.0	20.0	32.0	42.0	51.0	58.8	65.0	69.6

Table 4.3 Velocity components and cumulative distances at the beginning of each one second time interval

From Table 4.3 or Figure 4.3 the approximate time to the vertex is 2 seconds, using criterion (i). Hence the approximate maximum height reached is $(34.6 + 11.0) \times 1 = 45.6$ metres.

Using Table 4.3 and criterion (ii) the approximate time of flight is seen to be 6 seconds, and the corresponding horizontal range is 65.0 metres.

In Table 4.3 and Figure 4.4 the magnitude of the drag effect is indicated at each stage. This effect is seen to be large early in the flight, decreases to a relatively small value just past the top of the flight path, and then increases as the projectile falls.

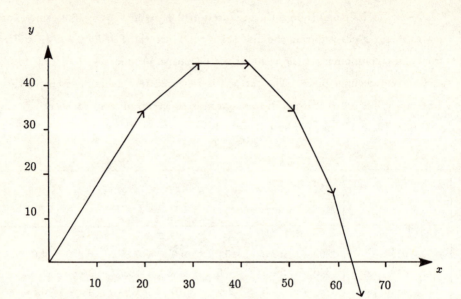

Figure 4.3 One second vector-addition approximation to the projectile's path

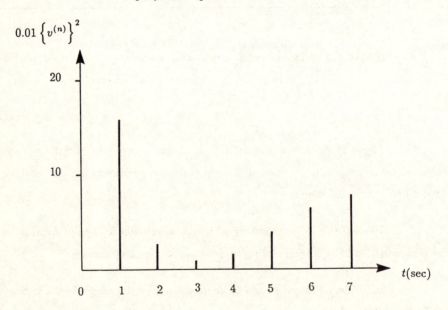

Figure 4.4 The magnitude of the approximate drag effect for each stage ($\times 10$)

The asymmetry of the flight path, including the fact that it is much steeper on the downward path of the flight, is clearly illustrated by the vector-addition approach illustrated in Figure 4.3. Comparison with the vector-addition diagram for the no-drag case under the same initial conditions clearly shows the reduction in range and maximum height caused by the drag effect (Figure 4.5).

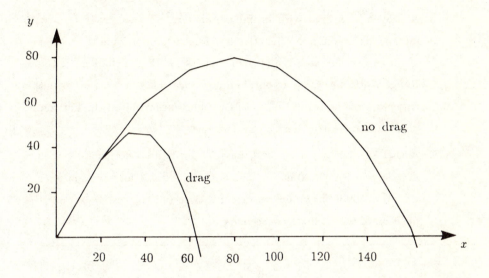

Figure 4.5 Comparison of vector-addition trajectories for drag and no-drag cases under the same initial conditions

The graphical technique can be used for any drag function, including those that are complicated expressions of v and y. It indicates the size of the drag effect at each step during its calculation and enables the angle of the projectile to be shown diagrammatically. The connection between the flight behaviour and the physical forces on the projectile is well illustrated by the technique. It is clear that the de Mestre - Catchpole technique has the potential for computer graphical treatment of many projectile problems. However it is only a first-order method as far as speed of convergence is concerned, and is not nearly as accurate as the modified Euler-Cauchy or the Runge-Kutta methods.

4.7 Exercises

1. Show that the density values given in Table 4.1 satisfy approximately the expression $\rho = \rho_0 exp(-0.0001036y)$. Obtain the percentage error at each specified height.

2. Fit a suitable curve to the temperature values given in Table 4.1.

3. A projectile is fired with initial parameters $v_0 = 40$ ms^{-1}, $\alpha = \pi/3$ radians in a medium in which the drag force is proportional to the square of the projectile's speed, and such that $\rho AC_D/(2m) = 0.01$ m^{-1}. Use a Runge-Kutta technique to obtain a numerical solution that predicts the trajectory of the projectile up to the point of impact on the horizontal plane. Choose a step-length of 0.1 seconds.

4. A shell of diameter 1 cm and mass 50 gm is fired horizontally with a muzzle velocity of 100 ms^{-1}. The G-function table for values of $v^2/100$ from 0 to 109 is shown below correct to 3 decimal places. Determine the position of the shell after 5 seconds.

$\dfrac{v^2}{100}$	0	1	2	3	4	5	6	7	8	9
0	.000	.002	.002	.003	.003	.004	.004	.005	.005	.005
10	.005	.006	.006	.006	.006	.006	.007	.007	.007	.007
20	.007	.008	.008	.008	.008	.008	.008	.008	.009	.009
30	.009	.009	.009	.009	.009	.009	.010	.010	.010	.010
40	.010	.010	.010	.010	.010	.011	.011	.011	.011	.011
50	.011	.011	.011	.011	.011	.012	.012	.012	.012	.012
60	.012	.012	.012	.012	.012	.012	.013	.013	.013	.013
70	.013	.013	.013	.013	.013	.013	.013	.013	.013	.014
80	.014	.014	.014	.014	.014	.014	.014	.014	.014	.014
90	.014	.014	.014	.014	.015	.015	.015	.015	.015	.015
100	.015	.015	.015	.015	.015	.015	.015	.015	.015	.015

5. Use the de Mestre - Catchpole graphical technique to determine the range and time of flight when $h = 0.5$ seconds, $v_0 = 40$ ms^{-1}, $\alpha = \pi/3$ radians for a projectile acted on by a body force proportional to the square of its speed with $\rho A C_D/(2m) = 0.01$ m^{-1}. (Compare your answers with those obtained from Exercise 3 above).

6. The long-range gun used by the Germans during World War I to shell Paris had a muzzle velocity of 1500 ms^{-1} and was fired at a fixed angle of 50° to the horizontal to reach the upper atmosphere where the air was considerably less dense. Suppose the mass of the shell was 45 kg, its diameter was 0.25 metres and the drag coefficient $C_D = 0.0002v$, where v is the speed of the shell at any instant. Determine the range of the gun.

7. For a cricket ball thrown, without spin at 30 ms^{-1} from a height of 2 metres, in still air which offers a resistance proportional to v^2, show that $E = 0.00295$ correct to three significant figures. (The mass of a cricket ball is 160 grams, its diameter is 7 cm and $C_D = 0.2$ at this speed).
Use numerical integration of the differential system (4.3) to show that the maximum range is achieved at approximately 42.5° to the horizontal. Show also that from 40° to 45° the range is within 50 cm of the maximum. Evaluate these differences as percentages of the maximum range.

8. Use the Ballistic Table 4.2 to determine the initial speed which must be imparted to a soccer ball projected at 48° to the horizontal so as to produce a 50 m kick.

9. A golf ball is driven off the tee at a speed of 61 ms^{-1} and an angle of 20° to the horizontal. If $C_D = 0.2$ while m/d^2 is 27 kg m^{-2} use the Ballistic Table 4.2 to determine the range to the first bounce if the ball does not spin.

10. With equations (4.3) recast in terms of the variable p obtain a set of equivalent equations for $p \equiv v_x$.

11. Oblique co-ordinates (L, D) were introduced into ballistics by K. Popoff. The L-axis is tangent to the trajectory at the initial position of the projectile, while the D-axis is vertically down from the L-axis to the

projectile. Show that for these co-ordinates

$$L = x \sec \alpha, \qquad D = x \tan \alpha - y,$$

and that equations (4.3) become

$$\frac{dv_L}{dt} = -Ev_L$$

$$\frac{dv_D}{dt} = -Ev_D + g$$

where $v_L = dL/dt$, $v_D = dD/dt$. (This form is useful for dive-bombing calculations.)

12. A gumboot-throwing contest is held on a day without wind at a horizontal ground. One contestant releases the gumboot from a height of 2 metres with an initial speed 30 ms^{-1} at an angle $\pi/6$ radians to the horizontal. Air is assumed to exert a drag force per unit mass equal to kv^2 with $k = 0.02$ m^{-1} and v is the speed of the centre of mass of the gumboot at any instant during its flight.

Use the graphical technique developed by de Mestre and Catchpole with time-step $h = 0.5$ to determine the approximate distance of the throw.

5. SMALL DRAG OR SMALL GRAVITY

"Then out of his pocket he fetched a stone,

And pelted it over the silent stream $----$"

A.B. (Banjo) Paterson (1864 - 1941)

5.1 Perturbation Techniques

Although the basic projectile equations (4.2) are usually solved by numerical methods it is possible to obtain closed-form approximate solutions when the ratio of drag force to gravity force is either very small or very large. For these cases a perturbation expansion in terms of a small parameter can be used to solve the governing differential equations to any order of the parameter desired. This was partly introduced in Chapter 2, Sections 2.3 and 2.4, for algebraic equations.

Before the technique is applied to projectile problems its application is illustrated first on a single differential equation and then on a simple system of differential equations.

Example 5.1   (A single differential equation)

Solve, for $t \geq 0$,

$$\frac{dx}{dt} = \epsilon x^2$$

with $x = -1$ when $t = 0$, where ϵ is a small positive parameter whose magnitude is much smaller than unity.

Solution

This non-linear problem can be solved directly by separation of the variables and integration to yield

$$x(t; \epsilon) = \frac{-1}{(1 + \epsilon t)}$$

Since $t > 0$ this solution is valid for all ϵ, including ϵ small. However when $\epsilon t < 1$ (i.e. for t values from 0 up to but not including ϵ^{-1}) the solution may be expanded as

$$x(t; \epsilon) = -1 + \epsilon t - \epsilon^2 t^2 + \cdots + (-1)^{n+1} \epsilon^n t^n + O(\epsilon^{n+1})$$

to any number of terms as required.

Although this expansion has been obtained directly a perturbation procedure is now illustrated which achieves the same result.

Consider the formal expansion

$$x(t; \epsilon) = x_0(t) + \epsilon x_1(t) + \epsilon^2 x_2(t) + O(\epsilon^3) \tag{5.1}$$

where $n = 2$ has been chosen in this case as providing enough terms to indicate the technique, and $x_i(t)$ are functions yet to be determined.

The initial condition will then be

$$-1 = x_0(0) + \epsilon x_1(0) + \epsilon^2 x_2(0) + O(\epsilon^3)$$

If terms of the same power of ϵ are equated the initial condition yields

$$x_0(0) = -1, \quad x_j(0) = 0 \quad (j = 1, 2) \tag{5.2}$$

The perturbation expansion (5.1) is substituted into the differential equation and yields

$$\frac{dx_0}{dt} + \epsilon \frac{dx_1}{dt} + \epsilon^2 \frac{dx_2}{dt} + O(\epsilon^3) = \epsilon \left[x_0 + \epsilon x_1 + O(\epsilon^2) \right]^2$$

$$= \epsilon \left[x_0^2 + 2\epsilon x_0 x_1 + O(\epsilon^2) \right]$$

If terms with the same powers of ϵ are equated then

$$\frac{dx_0}{dt} = 0, \quad \frac{dx_1}{dt} = x_0^2, \quad \frac{dx_2}{dt} = 2x_0 x_1 \tag{5.3}$$

Equations (5.3) and conditions (5.2) can be regrouped into the following sets of differential problems

$$\frac{dx_0}{dt} = 0, \qquad\qquad x_0(0) = -1 \tag{5.4}$$

$$\frac{dx_1}{dt} = x_0^2, \qquad\qquad x_1(0) = 0 \tag{5.5}$$

$$\frac{dx_2}{dt} = 2x_0 x_1, \qquad\qquad x_2(0) = 0 \tag{5.6}$$

The set (5.4) has the solutions $x_0(t) = -1$, and so the set (5.5) becomes

$$\frac{dx_1}{dt} = 1, \quad x_1(0) = 0$$

The solution is $x_1(t) = t$ which changes the set (5.6) into

$$\frac{dx_2}{dt} = -2t, \quad x_2(0) = 0$$

Thus $x_2(t) = -t^2$ and the approximate solution (5.1) becomes

$$x = -1 + \epsilon t - \epsilon^2 t^2 + O(\epsilon^3)$$

which is identical with the expanded solution obtained previously.

Example 5.2 (A system of differential equations)

When ϵ is a small parameter solve

$$\frac{dx}{dt} = -\epsilon y \tag{5.7}$$

$$\frac{dy}{dt} = \epsilon x + 1 \tag{5.8}$$

subject to $x = 2 - 1/\epsilon$, $y = 3$ when $t = 0$.

Solution

The system can be solved directly. First of all equation (5.7) is differentiated with respect to t and equation (5.8) used to produce

$$\frac{d^2 x}{dt^2} + \epsilon^2 x = -\epsilon$$

The general solution of this is

$$x = A \cos \epsilon t + B \sin \epsilon t - \frac{1}{\epsilon}$$

and equation (5.7) yields

$$y = A \sin \epsilon t - B \cos \epsilon t$$

Now the initial conditions give $A = 2$, $B = -3$, so the full solution is

$$x = 2\cos \epsilon t - 3\sin \epsilon t - \frac{1}{\epsilon}$$

$$y = 2\sin \epsilon t + 3\cos \epsilon t$$

For $\epsilon t < 1$ these can be expanded to yield

$$\left.\begin{aligned} x &= -\frac{1}{\epsilon} + 2 - 3\epsilon t - \epsilon^2 t^2 + O(\epsilon^3) \\[2mm] y &= 3 + 2\epsilon t - \frac{3}{2}\epsilon^2 t^2 + O(\epsilon^3) \end{aligned}\right\} \tag{5.9}$$

To be able to reproduce these results by a perturbation procedure it is necessary to make the correct formal expansions for x and y at the beginning. The form of the initial conditions suggests that solutions are sought of the form

$$x = \frac{x_{-1}(t)}{\epsilon} + x_0(t) + \epsilon x_1(t) + \epsilon^2 x_2(t) + O(\epsilon^3)$$

$$y = y_0(t) + \epsilon y_1(t) + \epsilon^2 y_2(t) + O(\epsilon^3)$$

provided $\epsilon \ll 1$.

The initial conditions become

$$x_{-1}(0) = -1, \quad x_0(0) = 2, \quad x_j(0) = 0 \quad (j = 1, 2)$$

$$y_0(0) = 3, \quad y_j(0) = 0 \quad (j = 1, 2)$$

while the equations (5.7) and (5.8) under the perturbation procedure yield

$$\frac{dx_{-1}}{dt} = 0, \quad \frac{dx_0}{dt} = 0, \quad \frac{dx_1}{dt} = -y_0, \quad \frac{dx_2}{dt} = -y_1$$

$$\frac{dy_0}{dt} = 1 + x_{-1}, \quad \frac{dy_1}{dt} = x_0, \quad \frac{dy_2}{dt} = x_1$$

Note that the differential equations have not uncoupled but that it is nevertheless possible to solve each one in turn so long as a logical back-and-forth procedure is adopted. The solutions are as expected

$$x_{-1}(t) = -1, \qquad x_0(t) = 2, \qquad x_1(t) = -3t, \qquad x_2(t) = -t^2$$

$$y_0(t) = 3, \qquad y_1(t) = 2t, \qquad y_2(t) = -\frac{3}{2}t^2$$

leading to the same perturbation expansion (5.9) as before.

5.2 Gravity or Drag Perturbations

The application of perturbation techniques to the basic projectile equations (4.2) will be illustrated by considering, in particular, the case in which the drag is proportional to the square of the speed. As a further simplification suppose that the drag coefficient C_D, the cross-sectional area A and the air density ρ are constant. For this case the normal and tangential components of the basic equations (4.2) become

$$\frac{dv}{dt} = -g \sin \psi - kv^2 \tag{5.10}$$

$$v\frac{d\psi}{dt} = -g \cos \psi \tag{5.11}$$

where ψ is still the angle between the tangent and the horizontal (and therefore non-dimensional) and $k = \rho A C_D/(2m)$. The initial conditions are

$$v = v_0, \qquad \psi = \alpha, \qquad x = 0, \qquad y = 0 \qquad \text{when} \qquad t = 0 \tag{5.12}$$

and x and y are defined by

$$\frac{dx}{dt} = v \cos \psi \tag{5.13}$$

$$\frac{dy}{dt} = v \sin \psi \tag{5.14}$$

Before these equations can be non-dimensionalised a representative speed and a representative time are needed. The initial speed v_0 of the projectile is obviously a representative speed. Although there is no obvious representative time, one can be constructed from either g or k which are dimensional constants within the problem.

Case 1   Using g, which has dimensions of length/(time)^2, a representative time v_0/g is constructed and so non-dimensional variables V and T are introduced by

$$v = v_0 V, \qquad\qquad t = \frac{v_0 T}{g}$$

The non-dimensional co-ordinates X, Y are introduced as before with

$$x = \frac{v_0^2 X}{g}, \qquad y = \frac{v_0^2 Y}{g}$$

The differential system (5.10) - (5.14) reduces to

$$
\left.
\begin{aligned}
\frac{dV}{dT} &= -\sin\psi - \epsilon V^2 \\[1ex]
V\frac{d\psi}{dT} &= -\cos\psi \\[1ex]
\frac{dX}{dT} &= V\cos\psi \\[1ex]
\frac{dY}{dT} &= V\sin\psi
\end{aligned}
\right\}
\qquad (5.15)
$$

$$\text{with} \quad V(0) = 1, \quad \psi(0) = \alpha, \quad X(0) = 0, \quad Y(0) = 0$$

where the parameter $\epsilon = k v_0^2/g$ measures the ratio of the initial drag force to the gravity force.

Case 2   Using k, which has dimensions $(\text{length})^{-1}$, a representative time $1/(k v_0)$ is constructed and non-dimensional variables are constructed by

$$v = v_0 V, \quad t = \frac{T}{k v_0}$$

The corresponding representative length is then $1/k$, so that

$$x = \frac{X}{k}, \quad y = \frac{Y}{k}$$

The differential system (5.10) - (5.14) reduces in this case to

$$
\left.
\begin{aligned}
\frac{dV}{dT} &= -V^2 - \bar{\epsilon}\sin\psi \\[2mm]
V\frac{d\psi}{dT} &= -\bar{\epsilon}\cos\psi \\[2mm]
\frac{dX}{dT} &= V\cos\psi \\[2mm]
\frac{dY}{dT} &= V\sin\psi \\[2mm]
\text{with } V(0) = 1, \quad \psi(0) &= \alpha, \quad X(0) = 0, \quad Y(0) = 0 \quad \text{and} \\[2mm]
\bar{\epsilon} = \frac{g}{kv_0^2} &= \frac{1}{\epsilon}
\end{aligned}
\right\}
\qquad (5.16)
$$

Thus $\bar{\epsilon}$ is the ratio of the gravity force to the initial drag force, which is the reciprocal of ϵ.

In some problems arising in athletics the drag effect can be expected to be much smaller than the gravity effect, so ϵ will be small and equations (5.15) can be used. In anti-aircraft gunnery on the other hand, the drag on a shell is much greater than its weight so $\bar{\epsilon}$ is small and equations (5.16) should be used.

Example 5.3   (Small drag problem)

A small kangaroo performs a jump with a take-off speed 15 ms^{-1} and a take-off angle $\pi/5$ radians. Air drag is assumed to be proportional to the square of the kangaroo's speed and the value of $\rho AC_D v_0^2/(2mg)$ is 0.1. Determine the horizontal distance moved by the kangaroo's centre of mass during the jump.

Solution

It is assumed that the kangaroo's centre of mass is at the same level on take-off as it is on landing.

If equations (5.15) are used the analysis could follow closely that presented in de Mestre (1986). However, the mathematics is simpler if the rectangular

Cartesian equivalent of equations (5.15) are used instead. These are

$$\left. \begin{array}{l} \dfrac{d^2 X}{dT^2} = -\epsilon \dfrac{dX}{dT} \left\{ \left[\dfrac{dX}{dT} \right]^2 + \left[\dfrac{dY}{dT} \right]^2 \right\}^{\frac{1}{2}} \\[4mm] \dfrac{d^2 Y}{dT^2} = -\epsilon \dfrac{dY}{dT} \left\{ \left[\dfrac{dX}{dT} \right]^2 + \left[\dfrac{dY}{dT} \right]^2 \right\}^{\frac{1}{2}} - 1 \end{array} \right\} \qquad (5.17)$$

with $dX/dT = \cos \pi/5$, $dY/dT = \sin \pi/5$, $X = 0$, $Y = 0$ when $T = 0$.

Now $\epsilon = 0.1$ and a solution is assumed of the form

$$X(T) = X_0(T) + \epsilon X_1(T) + O(\epsilon^2)$$

$$Y(T) = Y_0(T) + \epsilon Y_1(T) + O(\epsilon^2)$$

where X_0, Y_0 are essentially the no-drag solutions and X_1, Y_1 are the first-order corrections that include drag. When these expansions are substituted into equations (5.17), and terms of like order in powers of ϵ are equated, the following differential systems are obtained:

$$\left. \begin{array}{l} \dfrac{d^2 X_0}{dT^2} = 0 \\[4mm] \dfrac{d^2 Y_0}{dT^2} = -1 \end{array} \right\} \qquad (5.18)$$

with $dX_0/dT = \cos \pi/5$, $dY_0/dT = \sin \pi/5$, $X_0 = 0$, $Y_0 = 0$ when $T = 0$ and

$$\left. \begin{array}{l} \dfrac{d^2 X_1}{dT^2} = -\dfrac{dX_0}{dT} \left\{ \left[\dfrac{dX_0}{dT} \right]^2 + \left[\dfrac{dY_0}{dT} \right]^2 \right\}^{\frac{1}{2}} \\[4mm] \dfrac{d^2 Y_1}{dT^2} = -\dfrac{dY_0}{dT} \left\{ \left[\dfrac{dX_0}{dT} \right]^2 + \left[\dfrac{dY_0}{dT} \right]^2 \right\}^{\frac{1}{2}} \end{array} \right\} \qquad (5.19)$$

with $dX_1/dT = 0$, $dY_1/dT = 0$, $X_1 = 0$, $Y_1 = 0$ when $T = 0$. As expected the solution of equation (5.18) is the no-drag solution (1.4*) given by

$$X_0 = T \cos \frac{\pi}{5}$$

$$Y_0 = T \sin \frac{\pi}{5} - \frac{1}{2} T^2$$

When these are substituted into equations (5.19) they become

$$\frac{d^2 X_1}{dT^2} = -\cos\frac{\pi}{5}\left\{\cos^2\frac{\pi}{5} + \left[\sin\frac{\pi}{5} - T\right]^2\right\}^{\frac{1}{2}} \tag{5.20}$$

and

$$\frac{d^2 Y_1}{dT^2} = \left[T - \sin\frac{\pi}{5}\right]\left\{\cos^2\frac{\pi}{5} + \left[\sin\frac{\pi}{5} - T\right]^2\right\}^{\frac{1}{2}} \tag{5.21}$$

Integration of equation (5.21) twice with respect to T yields

$$Y_1(T) = \frac{1}{12}\left[T - \sin\frac{\pi}{5}\right]\left\{\cos^2\frac{\pi}{5} + \left[\sin\frac{\pi}{5} - T\right]^2\right\}^{\frac{3}{2}}$$

$$- \frac{1}{8}\cos^2\frac{\pi}{5}\left[T - \sin\frac{\pi}{5}\right]\left\{\cos^2\frac{\pi}{5} + \left[\sin\frac{\pi}{5} - T\right]^2\right\}^{\frac{1}{2}}$$

$$- \frac{1}{3}T + \frac{1}{12}\sin\frac{\pi}{5} - \frac{1}{8}\sin\frac{\pi}{5}\cos^2\frac{\pi}{5}$$

$$- \frac{1}{8}\cos^4\frac{\pi}{5}\ln\left[\frac{T - \sin\frac{\pi}{5} + \left\{\cos^2\frac{\pi}{5} + \left[\sin\frac{\pi}{5} - T\right]^2\right\}^{\frac{1}{2}}}{1 - \sin\frac{\pi}{5}}\right]$$

On landing, the kangaroo's centre of mass will be at $Y = 0$ which to order ϵ is given by

$$Y_0 + \epsilon Y_1 = O(\epsilon^2)$$

If the substitition

$$T = T_0 + \epsilon T_1 + O(\epsilon^2)$$

is made everywhere in this last equation, and terms with the same power of ϵ are equated, then

$$T_0\sin\frac{\pi}{5} - \frac{1}{2}T_0^2 = 0$$

and

$$T_1\sin\frac{\pi}{5} - T_0 T_1 + Y_1(T_0) = 0$$

Therefore, since the solution $T_0 = 0$ is associated with the take-off point, the time of flight is given approximately by $T_0 + \epsilon T_1$ where

$$T_0 = 2 \sin \frac{\pi}{5}$$

$$T_1 = -\frac{1}{2} - \frac{1}{4}\cos^2 \frac{\pi}{5} - \frac{\cos^4 \frac{\pi}{5}}{8 \sin \frac{\pi}{5}} \ln \left[\frac{1 + \sin \frac{\pi}{5}}{1 - \sin \frac{\pi}{5}} \right]$$

The horizontal distance travelled in this time is determined by integrating equations (5.20) twice with respect to T yielding

$$X_1(T) = -\frac{1}{6} \cos \frac{\pi}{5} \left\{ \cos^2 \frac{\pi}{5} + \left[\sin \frac{\pi}{5} - T \right]^2 \right\}^{\frac{3}{2}}$$

$$+ \frac{1}{2} \cos^3 \frac{\pi}{5} \left\{ \cos^2 \frac{\pi}{5} + \left[\sin \frac{\pi}{5} - T \right]^2 \right\}^{\frac{1}{2}}$$

$$- \frac{1}{2} T \cos \frac{\pi}{5} \sin \frac{\pi}{5} + \frac{1}{6} \cos \frac{\pi}{5} - \frac{1}{2} \cos^3 \frac{\pi}{5}$$

$$- \frac{1}{2} \cos^3 \frac{\pi}{5} \left[T - \sin \frac{\pi}{5} \right] \ln \left[\frac{T - \sin \alpha + \left\{ \cos^2 \alpha + \left[\sin \frac{\pi}{5} - T \right]^2 \right\}^{\frac{1}{2}}}{1 - \sin \frac{\pi}{5}} \right]$$

Substitution of the expressions for T_0 and T_1 into $X_0 + \epsilon X_1 + O(\epsilon^2)$ then produces

$$X = \sin \frac{2\pi}{5} - \epsilon \left\{ \frac{3}{2} \cos \frac{\pi}{5} - \frac{3}{4} \cos^3 \frac{\pi}{5} + \left[\frac{\cos^5 \frac{\pi}{5}}{8 \sin \frac{\pi}{5}} + \frac{1}{2} \cos^3 \frac{\pi}{5} \sin \frac{\pi}{5} \right] \ln \left[\frac{1 + \sin \frac{\pi}{5}}{1 - \sin \frac{\pi}{5}} \right] \right\}$$

$$= 0.84$$

remembering that the error is order $10^{-2} (= \epsilon^2)$. Since $x = v_0^2 X/g$ the kangaroo jumped a horizontal distance of 19 metres approximately.

Example 5.4   (Small gravity problem)

An Oerlikon anti-aircraft gun fires a 20 mm shell with a muzzle speed of 1050 ms^{-1} at an angle of 20° to the horizontal. The drag on the shell is proportional to the square of the speed, and the ratio of the initial drag force to the projectile's weight is 51. Determine the speed, the angle of inclination of the shell, its altitude and downrange at half-second intervals for $0 \leq t \leq 4$.

Solution

The analysis was originally produced by Milenski (1969).

Since $\bar{\epsilon} = 1/51 = 0.0196$ the centre of mass of the shell is governed by equations (5.16). For this example

$$\frac{dV}{dT} = -V^2 - \bar{\epsilon}\sin\psi$$

and if the second of (5.16) is multiplied by $\cos\psi$ it becomes

$$V\frac{d(\sin\psi)}{dT} = \bar{\epsilon}(\sin^2\psi - 1) \tag{5.23}$$

The formal expansions

$$V = V_0(T) + \bar{\epsilon}V_1(T) + 0(\bar{\epsilon}^2)$$

$$\sin\psi = \Psi_0(T) + \bar{\epsilon}\Psi_1(T) + O(\bar{\epsilon}^2)$$

are made where V_0 and Ψ_0 now represent the solutions for the straight-line motion of the shell from the gun through the resisting medium, while V_1 and Ψ_1 represent the first-order corrections to that motion due to gravity effects.

Substitution into the two differential equations and their initial conditions $V = 1$, $\psi = \pi/9$ when $T = 0$ yields the differential systems

$$\left.\begin{aligned} \frac{dV_0}{dT} &= -V_0^2 \\[2mm] V_0\frac{d\Psi_0}{dT} &= 0 \end{aligned}\right\} \tag{5.24}$$

with $V_0 = 1$, $\Psi_0 = \sin\pi/9$ when $T = 0$;

$$\left.\begin{aligned} \frac{dV_1}{dT} &= -\Psi_0 - 2V_0V_1 \\[2mm] V_0\frac{d\Psi_1}{dT} + V_1\frac{d\Psi_0}{dT} &= \Psi_0^2 - 1 \end{aligned}\right\} \tag{5.25}$$

with $V_1 = 0$, $\Psi_1 = 0$ when $T = 0$.

The solution of the system (5.24) is

$$V_0 = \frac{1}{1+T}, \qquad \Psi_0 = \sin\frac{\pi}{9}$$

When these are substituted into (5.25), the solution of the resulting system is

$$V_1 = \frac{1}{3} \sin \frac{\pi}{9} \left\{ \frac{1}{(1+T)^2} - (1+T) \right\}$$

$$\Psi_1 = \frac{1}{2} \cos^2 \frac{\pi}{9} \{1 - (1+T)^2\}$$

Therefore

$$V \approx \frac{1}{1+T} + 0.0022 \left\{ \frac{1}{(1+T)^2} - (1+T) \right\}$$

and

$$\sin \psi \approx 0.3420 + 0.0087 \{1 - (1+T)^2\}$$

When these are substituted into the third and fourth of equations (5.16), namely

$$\frac{dX}{dT} = V(1 - \sin^2 \psi)^{\frac{1}{2}}$$

and

$$\frac{dY}{dT} = V \sin \psi,$$

and the relevant integrations are performed the expressions obtained are

$$X \approx 0.9397 \ln(1+T) + 0.0005 \left\{ (1+T)^2 + 3 - \frac{4}{(1+T)} - 6 \ln(1+T) \right\}$$

$$Y \approx 0.3420 \ln(1+T) + 0.0004 \left\{ -2(1+T)^2 + 4 - \frac{2}{(1+T)} + 2 \ln(1+T) \right\}$$

These could also be obtained by considering the Cartesian co-ordinate version of equations (5.16), but the analysis would not be as efficient in this example as it was in Example 5.3.

When the expressions for X, Y and V are converted back to dimensional form it is possible to construct Table 5.1 for the speed v, angle of inclination ψ, altitude y and downrange x at half-second steps over the time interval $0 \le t \le 4$.

In both these examples the problem could have been solved semi-analytically by the Bernoulli technique of Chapter 3. However this would not produce explicit expressions for $v(t)$, $\psi(t)$, $x(t)$ and $y(t)$. This is also true for problems with drag proportional to other powers of v. For more complicated drag expressions the Bernoulli technique is no longer applicable whereas the perturbation technique may still be useful in many cases.

Time(s)	Speed(ms^{-1})	Angle(degrees)	Altitude(m)	Downrange(m)
0.0	1050	20.0	0	0
0.5	847	19.7	160	443
1.0	709	19.4	289	807
1.5	609	19.0	397	1117
2.0	534	18.5	489	1387
2.5	475	18.0	568	1626
3.0	427	17.4	636	1840
3.5	388	16.8	696	2035
4.0	355	16.1	748	2213

Table 5.1 Flight characteristics for an Oerlikon shell during the first 4 seconds

5.3 Exercises

1. Use a perturbation procedure to solve

$$\frac{dx}{dt} = 1 + \epsilon x$$

with $x = -1$ when $t = 0$, where ϵ is a small positive parameter. Find the solution correct to $O(\epsilon^2)$.

2. Use a perturbation procedure to solve the system

$$6\frac{dx}{dt} + 2\frac{dy}{dt} - \epsilon y + \epsilon^2 = 0,$$

$$3\frac{dx}{dt} - \frac{dy}{dt} + \epsilon x + 1 = 0$$

with $x = 1$, $y = 2 + \epsilon$ when $t = 0$ and ϵ is a small positive parameter. Solve correct to $O(\epsilon^3)$.

3. Re-do the analysis for Example 5.3 when the governing equations are (5.15).

4. For a certain female long-jumper on whom the air drag is proportional to the square of her speed, the value of ϵ is 0.06. Obtain an expression for the horizontal distance moved by her centre of mass when her vertical distance has reached a maximum.

5. Suppose that a projectile moves through a medium where the drag force is $\frac{1}{2}\rho v^n C_D A$ with ρ, C_D and A different constants and n a positive integer. Derive the normal and tangential basic equations for its flight in terms of a parameter ϵ which defines the ratio of the initial drag force to the gravity force. Solve the equations correct to first order in ϵ when $\epsilon \ll 1$.

6. Re-do the analysis for Example 5.4 when the governing equations are the Cartesian co-ordinate version of (5.16).

7. A quantity of considerable interest in anti-aircraft gunnery is the gravity drop defined in the non-dimensional terms of Example 5.4 as $Y(T;0) - Y(T;\bar{\epsilon})$. Show that, for an initial angle of firing α, the gravity drop is

$$\bar{\epsilon}\left[\frac{\cos^2 \alpha}{4} \left\{ (1+T)^2 - 1 - 2\ln(1+T) \right\} \right.$$

$$\left. + \frac{\sin^2 \alpha}{6} \left\{ (1+T)^2 - 3 + \frac{2}{1+T} \right\} \right] + O(\bar{\epsilon}^2)$$

8. Solve the small gravity problem illustrated in Example 5.4 when the drag on the shell is proportional to v^n, where n is a positive number. Check your answer by verifying that when $n = 1$ the pertinent results of Chapter 2 are recovered.

9. With $n = 2$ and $\epsilon = kv_0^2/g$ the Bernoulli equation in Chapter 3 becomes

$$\frac{dV}{d\psi} - V \tan \psi = \epsilon V^3 \sec \psi$$

when reduced to non-dimensional form. Applying the usual initial conditions $\mathbf{R} = \mathbf{0}, \mathbf{V} = \hat{\mathbf{v}}_0$, solve this by a perturbation expansion to obtain the zeroth and first order approximations to $V(\psi; \epsilon)$ for ϵ small . Obtain the corresponding approximation for the maximum height reached and the time taken to reach this height.

10. Using intervals of $10°$ show that, as α changes from $10°$ to $70°$ for a long jumper with take-off speed 9 ms^{-1}, the distance jumped varies by about 3 to 4 metres. What appears to be the best take-off angle to achieve maximum horizontal distance?

11. For a projectile fired at an angle α from a dimensionless height H in a medium whose drag is proportional to v^n the horizontal distance travelled is given approximately by

$$X = X_0(\alpha; H) + \epsilon X_1(\alpha; H)$$

where ϵ is small and replaces kv_0^n/g. Show that the optimum angle for maximum horizontal distance obtained from this height is given by

$$\alpha_0 - \epsilon \left\{ \frac{dX_1(\alpha_0; H)/d\alpha}{d^2 X_0(\alpha_0; H)/d\alpha^2} \right\}$$

where

$$\sin \alpha_0 = \frac{1}{2(1 + H)^{\frac{1}{2}}}$$

12. When a spear is fired underwater from a speargun the drag can be considered to be proportional to the square of the speed, and is much greater than the gravity effect. For a diver endeavouring to spear a fish he usually releases the spear horizontally after aiming at the centre of the fish. If the fish is $2h$ cm high and the initial spear speed is v_0 cms^{-1} determine the distance of firing beyond which the fish will get away.

6. CORRECTIONS DUE TO OTHER EFFECTS

"A falling body ought by reason of the earth's diurnal motion
to advance eastward and not fall to the west as the vulgar opinion is."

Sir Isaac Newton (1642-1727)

6.1 Effects Other Than Constant Gravity and Variable Drag

The basic equations (4.1) or (4.3) for a projectile are essentially restricted to two dimensions. If an OZ axis is introduced orthogonal to the OX and OY axes then the absence of forces in the z-direction combined with the initial conditions $z = dz/dt = 0$ when $t = 0$ ensures that $z \equiv 0$ for all t. However, any non-symmetrical variation (such as spin or yaw) in the z-direction will produce a drift effect on the projectile in this lateral direction.

As soon as spin is imparted to a projectile to reduce yaw and impose stability of flight there are many effects such as cross-forces and cross-torques which have to be included in the equations of motion (see Chapter 7). Even in the absence of spin the non-symmetrical shape of some projectiles or the yawing of a symmetrical projectile will give rise to a cross-force known as lift.

The spinning of the earth also requires the inclusion of pseudo-forces (Coriolis and centrifugal) in the governing equation, particularly when the projectile has a very long flight path. For these problems gravity variations due to altitude and the non-spherical shape of the earth may also be important.

Variations in the density and the speed of sound due to changes in temperature and humidity from standard conditions will also produce secondary effects. But the major meteorological effect for most projectiles is the wind, which produces a considerable alteration to the drag force.

Finally the shape of a projectile, combined with the fact that the resultant of all the forces acting on it may not pass through its centre of gravity, ensures that there will usually be an overturning moment. This contributes to the rotation of a projectile in flight known as the yaw effect which, if not controlled, greatly increases the drag and therefore reduces the range. These extra effects are investigated individually, and then at the end of the chapter it is considered how they might be treated collectively.

6.2 Coriolis Corrections

Newton's laws of motion should only be applied to inertial frames of reference, that is, frames that are fixed at rest or are moving with constant velocity. When the frame of reference is rotating (as for a frame set at the projection point of a projectile from the Earth's surface) Coriolis corrections are needed because the reference frame is non-Newtonian or non-inertial.

Consider a frame of reference F centred at O on the Earth's surface and rotating with angular speed ω relative to a fixed frame of reference F' centred at O' at the centre of the Earth, which is a fixed distance from O. The velocity of a particle in the frame F' is the vectorial sum of the velocity of the particle in frame F and the velocity due to the rotation of frame F relative to frame F' (see Christie, 1964). Thus

$$\frac{d\mathbf{r}'}{dt'} = \frac{d\mathbf{r}}{dt} + \omega \times \mathbf{r}$$

where $d\mathbf{r}'/dt'$ is the particle's velocity measured in frame F' and $d\mathbf{r}/dt$ is the particle's velocity measured in frame F. Therefore (assuming $dt'/dt = 1$)

$$\frac{d^2\mathbf{r}'}{dt'^2} = \frac{d}{dt'}\left\{ \frac{d\mathbf{r}}{dt} + \omega \times \mathbf{r} \right\}$$

$$= \frac{d}{dt}\left\{ \frac{d\mathbf{r}}{dt} + \omega \times \mathbf{r} \right\} + \omega \times \left\{ \frac{d\mathbf{r}}{dt} + \omega \times \mathbf{r} \right\}$$

$$= \frac{d^2\mathbf{r}}{dt^2} + \frac{d\omega}{dt} \times \mathbf{r} + 2\omega \times \frac{d\mathbf{r}}{dt} + \omega \times (\omega \times \mathbf{r}) \qquad (6.1)$$

The angular velocity of the Earth ω can be considered as approximately constant, and so $d\omega/dt$ can be approximated to zero. Consequently the Euler term $d\omega/dt \times \mathbf{r}$ can be neglected in these considerations.

The angular speed of the Earth is $2\pi/(24\times3600) = 7.27\times10^{-5}$ sec^{-1}, and so ω^2 is very small. Therefore the centrifugal term $\omega \times (\omega \times \mathbf{r})$ will have an extremely small magnitude compared with those of the other terms and is also neglected when dealing with projectiles near the Earth's surface.

The remaining correction term $2\boldsymbol{\omega} \times d\mathbf{r}/dt$, when multiplied by the mass of the projectile and transferred to the other side of the momentum equation, becomes $-2m\boldsymbol{\omega} \times \mathbf{v}$. This term is then treated as a fictitious force called the Coriolis force. Its effect is quite significant for such widely differing motions as those associated with Foucault's pendulum or the trajectory of an intercontinental ballistic missile.

To illustrate the Coriolis effect a non-spinning projectile is investigated under the influence of gravity and Coriolis force, but with negligible drag effect. Then the gravitational force is the only true force acting and from equation (6.1)

$$\frac{d^2\mathbf{r}}{dt^2} + 2\boldsymbol{\omega} \times \mathbf{v} = \mathbf{g} \tag{6.2}$$

with the usual initial conditions $\mathbf{v} = \mathbf{v}_0$, $\mathbf{r} = \mathbf{0}$ when $t = 0$. Equation (6.2) is rewritten as

$$\frac{d^2\mathbf{r}}{dt^2} = \mathbf{g} - 2\boldsymbol{\omega} \times \frac{d\mathbf{r}}{dt}$$

and integrated with respect to t yielding

$$\mathbf{v} = \mathbf{v}_0 + \mathbf{g}t - 2\boldsymbol{\omega} \times \mathbf{r}$$

This can be substituted back into equation (6.2) to yield

$$\frac{d^2\mathbf{r}}{dt^2} = \mathbf{g} - 2\boldsymbol{\omega} \times \mathbf{v}_0 - 2\boldsymbol{\omega} \times \mathbf{g}t + 4\boldsymbol{\omega} \times (\boldsymbol{\omega} \times \mathbf{r})$$

But expressions like the last term have already been neglected when going from equation (6.1) to equation (6.2) and therefore will be neglected again for consistency. Integration of the remaining terms twice with respect to t produces

$$\mathbf{v} = \mathbf{v}_0 + \mathbf{g}t - 2\boldsymbol{\omega} \times \left(\mathbf{v}_0 t + \frac{1}{2}\mathbf{g}t^2 \right) \tag{6.3}$$

and

$$\mathbf{r} = \mathbf{v}_0 t + \frac{1}{2}\mathbf{g}t^2 - \boldsymbol{\omega} \times \left(\mathbf{v}_0 t^2 + \frac{1}{3}\mathbf{g}t^3 \right) \tag{6.4}$$

The expression for $\mathbf{r}$ has a term containing t^3 which may be significant for large flight times even though ω is small.

Suppose that λ is the astronomical latitude of a point O on the Earth's surface with respect to a point O' at the centre of the Earth. Suppose also that OY is the direction of the vertical at O, OZ is parallel to the North direction at O and OX is parallel to the West direction at O (see Figure 6.1).

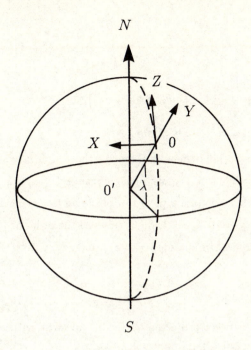

Figure 6.1 : Astronomical latitude and co-ordinate axes at O

The components of the various vector quantities are

$$\boldsymbol{\omega} = \omega[0, \ \sin \lambda, \ \cos \lambda]$$

$$\mathbf{g} = [0, \ -g, \ 0]$$

$$\mathbf{v}_0 = [v_{0x}, \ v_{0y}, \ v_{0z}]$$

$$\mathbf{r} = [x, \ y, \ z]$$

where λ is negative in the Southern Hemisphere. Equations (6.3) and (6.4) can then be written as

$$\left. \begin{aligned} v_x &= v_{0x} - 2\omega(v_{0z} \sin \lambda - v_{oy} \cos \lambda)t - \omega g t^2 \cos \lambda \\[2mm] v_y &= v_{0y} - gt - 2\omega v_{0x} t \cos \lambda, \\[2mm] v_z &= v_{0z} + 2\omega \, v_{0x} t \sin \lambda \end{aligned} \right\} \tag{6.5}$$

and

$$
\left.
\begin{aligned}
x &= v_{0x}t - \omega(v_{0z}\sin\lambda - v_{0y}\cos\lambda)t^2 - \frac{1}{3}\omega g t^3 \cos\lambda \\
y &= v_{0y}t - \left(\omega v_{0x}\cos\lambda + \frac{1}{2}g\right)t^2 \\
z &= v_{0z}t + \omega v_{0x}t^2 \sin\lambda
\end{aligned}
\right\}
\tag{6.6}
$$

The Coriolis effect is given by the terms in equations (6.5) and (6.6) which contain ω. Thus it is only of significance in aiming a projectile when t is relatively large. The ratio of a typical Coriolis term to a typical gravity term is $O(\omega t)$, and since ω is of order 10^{-4} sec^{-1}, flight times of the order of 1000 seconds are needed before there is a deviation amounting to approximately 10% of the range.

Example 6.1

For a body projected vertically upwards with speed v_0 show that its deflection on reaching the ground is $4\omega v_0^3 \cos\lambda/(3g^2)$ to the West. When this body reaches the top of its trajectory a second body is released from rest at the same height. Determine the deflection of the second body, when it too reaches the ground. Assume that the only effects are gravity and Coriolis.

Solution

For the first body the time of flight to the ground is given by $y = 0$. Therefore equation (6.6) with $\mathbf{v}_0 = [0,\ v_0,\ 0]$ yields

$$
0 = v_0 t_f - \left(0 + \frac{1}{2}g\right)t_f^2
$$

Thus $t_f = 2v_0/g$ which shows that the Coriolis effect does not influence the time of flight in this example. Substitution into the x-component of equation (6.6) produces

$$
x = \omega v_0 \cos\lambda \frac{4v_0^2}{g^2} - \frac{1}{3}\omega g \frac{8v_0^3}{g^3}\cos\lambda
$$
$$
= \frac{4\omega v_0^3 \cos\lambda}{3g^2}
$$

while the z-component is zero. Therefore the deflection is $4\omega v_0^3 \cos \lambda/(3g^2)$ to the West, and is only zero at the poles where $\cos \lambda = 0$. At the equator $\cos \lambda = 1$ and there is a maximum effect.

The first body reaches the top of its trajectory when $v_y = 0$. This gives $t_m = v_0/g$ and its height is obtained from the y-component of equation (6.6) as $v_0^2/(2g)$.

The second body with initial velocity $[0, 0, 0]$ therefore falls a vertical distance $v_0^2/(2g)$. Thus from the y-component of equation (6.6) its time of flight is v_0/g also, and its position vector has a zero z-component and an x-component given by

$$x = -\frac{1}{3}\omega g \frac{v_0^3}{g^3} \cos \lambda$$

Its deflection is therefore $\omega v_0^3 \cos \lambda/(3g^2)$ to the East.

This example illustrates one of the peculiarities of the Coriolis effect in that it seems to produce deflections in different directions from almost identical situations. However, the situations are not identical and it is left to the reader to sort out the apparent paradox.

6.3 Gravity Corrections

Each body in the universe will exert some force on a projectile. However, for a projectile near the surface of the Earth it has been calculated that the moon has an effect of the order of 10^{-7} times that of the Earth's effect, while the sun and the planets have an even smaller effect.

Although the gravitational effect of the Earth was given in Chapter 1 there are two perturbation corrections that should be made for projectiles near the Earth's surface. The first is an altitude correction, and the second takes account of the slightly non-spherical shape of the Earth which causes different values of g at different latitudes.

Case 1   Altitude correction

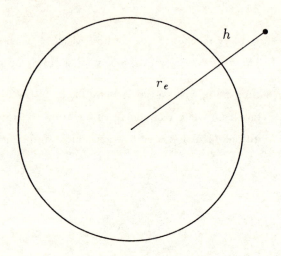

Figure 6.2 Projectile at height h above the Earth's surface

If h is the altitude of a projectile above the spherically symmetric Earth's surface (Figure 6.2), then equation (1.1) can be rewritten as

$$\mathbf{F} = -\frac{Gm_e m}{(r_e + h)^2}\hat{\mathbf{j}}$$

$$= -\frac{Gm_e m}{r_e^2}\left(1 + \frac{h}{r_e}\right)^{-2}\hat{\mathbf{j}}$$

where $r_e (= 6380\text{km})$ is the radius of the Earth as defined in Chapter 1. Provided that $h \ll r_e$ a binomial expansion can be applied and, if only the first two terms are retained, the main gravity term and an approximate altitude correction is obtained. Therefore, the gravitational force per unit mass is

$$\frac{\mathbf{F}}{m} = -g\left\{1 - \frac{2h}{r_e} + O\left(\frac{h^2}{r_e^2}\right)\right\}\hat{\mathbf{j}} \tag{6.7}$$

which predicts that the gravity effect is reduced at higher altitudes. This formula confirms the statement in Chapter 1 that the altitude effect is less than 1% for projectiles within 30 km of the Earth's surface.

Case 2   Non-spherical or latitude correction

The non-spherical correction is not as easily derived as the altitude correction. The Earth does not have the exact shape of a sphere, but is well approximated by an oblate spheroid with its polar axis as the minor axis. The assumption is made that the material of the Earth is approximately constant with density ρ. It is required to determine the gravitational force per unit mass at any point O on the Earth's surface where the general astronomical latitude is denoted by λ. The derivation of the gravitation correction due to latitude will be based on that given by Moulton (1962).

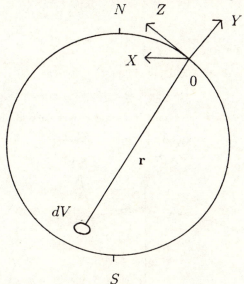

Figure 6.3 A volume element inside the earth

The attractive force exerted on a projectile of unit mass at O by a volume element dV inside the Earth is $(G\rho/r^2)dV\ \hat{\mathbf{r}}$ (see Figure 6.3). The component of this force per unit mass in the $\hat{\mathbf{j}}$ direction is therefore $-(G\rho y/r^3)dV\ \hat{\mathbf{j}}$, and consequently for the whole Earth the gravitational force per unit mass at O in the OY direction is given by

$$\frac{\mathbf{F}\cdot\hat{\mathbf{j}}}{m} = -G\rho \int_{\text{Earth}} \frac{y\,dV}{r^3}$$

Since the shape of the Earth is approximately a sphere a spherical polar transformation is used of the form

$$x = r \cos \theta \, \sin \phi, \quad y = r \sin \theta, \quad z = -r \cos \theta \, \cos \phi$$

with

$$dV = r^2 \cos \theta \, dr \, d\theta \, d\phi$$

Thus the vertical component of the gravitational force per unit mass at O is

$$\frac{\mathbf{F} \cdot \mathbf{\hat{j}}}{m} = -G\rho \int_0^{2\pi} \int_{-\frac{\pi}{2}}^{0} \int_0^{r_s} \sin \theta \, \cos \theta \, dr \, d\theta \, d\phi$$

$$= -G\rho \int_0^{2\pi} \int_{-\frac{\pi}{2}}^{0} r_s \, \sin \theta \, \cos \theta \, d\theta \, d\phi \qquad (6.8)$$

where $r_s = r_s(\theta, \phi)$ is the spherical polar equation for the spheroidal surface of the Earth with O as origin. Once the form of $r_s(\theta, \phi)$ is known it can be substituted into equation (6.8). When the necessary integrations are carried out the gravitational correction due to latitude is obtained.

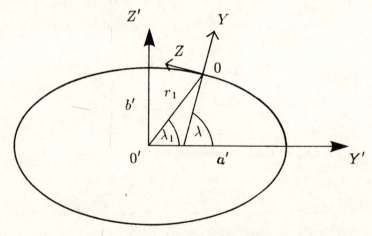

Figure 6.4 The Earth section (highly exaggerated) and the coordinate systems in the plane containing O and the polar axis $O'Z'$

Clearly the equation of the Earth's surface with the centre O' of the spheroid as origin is

$$\left[\frac{x'}{a'}\right]^2 + \left[\frac{y'}{a'}\right]^2 + \left[\frac{z'}{b'}\right]^2 = 1$$

where the polar axis is denoted by $O'Z'$. Since $b' < a'$ for an oblate spheroid then

$$(b')^2 = (a')^2(1 - e^2)$$

where e denotes the eccentricity of the meridional section depicted in Figure 6.4. But $e^2 = 0.0067$ for the Earth, and so terms of order e^4 and higher can be neglected. The equation for the surface of the Earth can then be written approximately as

$$(x')^2 + (y')^2 + (1 + e^2)(z')^2 = (a')^2 \tag{6.9}$$

This equation is now converted to a polar equation with O as origin. To do this the transformation

$$x' = x$$

$$y' = y \cos \lambda - z \sin \lambda + r_1 \cos \lambda_1$$

$$z' = y \sin \lambda + z \cos \lambda + r_1 \sin \lambda_1$$

is introduced where (r_1, λ_1) are the polar co-ordinates of O with respect to O' as shown in Figure 6.4. This transformation corresponds to a translation from O' to O and a rotation of the axes about OX through an angle λ.

Approximations to $r_1 \cos \lambda_1$ and $r_1 \sin \lambda_1$ can be obtained by noting that $(x', y', z') = (0, \ r_1 \cos \lambda_1, \ r_1 \sin \lambda_1)$ at O. Therefore from equation (6.9)

$$r_1 = a' \left\{ 1 - \frac{1}{2} e^2 \sin^2 \lambda_1 + O(e^4) \right\}$$

for small e^2. Also since $\cos \lambda$ is the direction cosine that the normal OY at O makes with the $O'Y'$ direction then

$$\cos \lambda = \frac{y'}{\sqrt{(x')^2 + (y')^2 + (1 + e^2)^2(z')^2}}$$

evaluated at O. Thus for small e^2

$$\cos \lambda = \cos \lambda_1 \left\{ 1 - e^2 \sin^2 \lambda_1 + O(e^4) \right\}$$

which on rearranging leads to

$$\cos \lambda_1 = \cos \lambda \left\{ 1 + e^2 \sin^2 \lambda + O(e^4) \right\}$$

and

$$\sin \lambda_1 = \sin \lambda \left\{ 1 - e^2 \cos^2 \lambda + O(e^4) \right\}$$

On substitution into equation (6.9) the polar equation of the surface of the Earth is therefore given approximately by

$$r_s^2 + e^2 r_s^2 (\sin \lambda \ \sin \theta - \cos \lambda \ \cos \theta \ \cos \phi)^2 + 2 r_s a' \left(1 + \frac{1}{2} e^2 \sin^2 \lambda \right) \sin \theta = 0$$

One solution is $r_s = 0$ corresponding to the point O, while the other gives the approximate equation of the Earth's surface as

$$r_s = -2a' \left\{ 1 + \frac{1}{2} e^2 \sin^2 \lambda - e^2 (\sin \lambda \ \sin \theta - \cos \lambda \ \cos \theta \ \sin \phi)^2 \right\} \sin \theta.$$

When this is substituted into equation (6.8) and the integrations carried out it yields

$$\frac{\mathbf{F} \cdot \hat{\mathbf{j}}}{m} = -\frac{4}{3} \pi \rho a' G \left\{ 1 - \frac{e^2}{5} \left(1 - \frac{1}{2} \sin^2 \lambda \right) + O(e^4) \right\}.$$

But the mass of the Earth is

$$m_e = \frac{4}{3} \pi \rho (a')^2 b'$$

$$= \frac{4}{3} \pi \rho r_e^3$$

where r_e is the geometric mean radius of the Earth in the sense that $r_e = \{(a')^2 b'\}^{\frac{1}{3}}$. Now $b' = a'(1 - e^2)^{\frac{1}{2}}$, therefore

$$a' = r_e \left\{ 1 + \frac{1}{6} e^2 + O(e^4) \right\}$$

Since $g = G m_e / r_e^2$ it is seen finally that

$$\frac{\mathbf{F} \cdot \hat{\mathbf{j}}}{m} = -g \left\{ 1 - \frac{e^2}{30} (1 - 3 \sin^2 \lambda) + O(e^4) \right\} \tag{6.10}$$

This gives the approximate correction for latitude due to the Earth being non-spherical. Note that the gravitational pull at the poles ($\lambda = \pm\pi/2$) is stronger than at the equator ($\lambda = 0$).

6.4 Density, Temperature, Pressure and Humidity Variations

A discussion of the effect of variations in density and temperature with altitude was given in Chapter 4, which included the relevant Standard Table 4.1. Sometimes these effects can be used to great advantage, as for example when a missile or projectile is fired into the upper atmosphere where the density of the air is significantly reduced. When this happens the drag force given by equation (3.1) is also reduced resulting in a greater range being obtained.

The density of the air varies at a particular point as the pressure, temperature and humidity change. The formula relating the first two of these to the density is the perfect gas equation

$$p = \rho R \tau$$

where p denotes the pressure in pascals, τ denotes the absolute temperature in degrees kelvin, ρ denotes the density in kg m^{-3}, and R is the gas constant, which for air has a value $0.287\,\text{kJ}\,\text{kg}^{-1}\text{K}^{-1}$. Thus a rise in pressure will produce an increase in density, but a rise in temperature decreases the density. Hence a projectile will fly more easily through the air on a hot day or when the barometric pressure is lower than on a cold day or when the pressure is higher. Pressure and temperature extremes at most places in Australia indicate that these could change the range of a football kick or cricket shot by up to 4 metres.

The formula relating density to humidity variations can also be obtained from the perfect gas equation above. Consider some moist air containing water vapour at pressure p_w and temperature τ while the air particles contribute a pressure p_a at the same temperature τ. The pressure of the moist air is then

$$p = p_a + p_w$$

Now

$$p_w = \rho_w R_w \tau$$

and

$$p_a = \rho_a R_a \tau$$

where R_w and R_a are the respective gas constants for water and air, with ρ_a and ρ_w denoting the corresponding density contributions. The density of the moist air (ρ_m) is thus given by

$$\rho_m = \rho_a + \rho_w$$

$$= \frac{p_w}{R_w \tau} + \frac{(p - p_w)}{R_a \tau}$$

Under the same conditions of temperature and pressure the density of dry air (ρ_d) would be

$$\rho_d = \frac{p}{R_a \tau}$$

and therefore

$$\frac{\rho_m}{\rho_d} = \frac{p - p_w \left(1 - \frac{R_a}{R_w}\right)}{p}$$

Now R_a/R_w has a value 0.62 and so the relation between the density of moist air and the density of dry air can be expressed as

$$\frac{\rho_m}{\rho_d} = 1 - 0.38 \left[\frac{p_w}{p}\right] \tag{6.11}$$

As the moisture increases, the density therefore decreases. Since the saturated vapour pressure of water increases with temperature the effect of humidity is likely to be more pronounced on a hot day. Calculations show that such variations due to humidity are unlikely to be more than 1%. It is interesting to note that when the comment is made in cricket that the ball swings well because of the heavy and humid atmosphere, equation (6.11) cannot support this. An atmosphere cannot be both humid and heavy since an increase in humidity implies a decrease in density.

6.5 Lift and Sideways Aerodynamic Corrections

For the major portion of its trajectory the longitudinal axis of a projectile will not coincide with the direction of the projectile's velocity. When this happens the angle of yaw between these two directions is not zero and the drag force may be considerably altered. For many shells in flight, the drag is doubled when the angle of yaw reaches about 13°. Although the projected cross-sectional area frequently increases with increasing yaw, it is more usual to take the non-yaw cross-sectional area A as a fixed representative area and consider C_D as varying with the angle of yaw δ. For an axisymmetric body, the variation in C_D must be an even function of δ, so for small δ the quadratic approximation is made without any linear term.

When the longitudinal axis of any symmetrical projectile coincides with its velocity direction, the angle of yaw is zero and the aerodynamic forces on any part of the projectile's surface are the mirror image of those on the part opposite. Thus the longitudinal components add to give a finite drag while the lateral components cancel to zero.

However, for any non-symmetrical projectile, or for a symmetrical projectile with non-zero yaw, the resultant of the aerodynamic forces on the projectile has components in the normal plane as well as its drag component in the tangential direction. This normal component tends to push the projectile at right angles to its direction of motion. The sideways part of this normal component contributes to the drift of the projectile. The remainder of this normal component is termed the lift force and lies in the vertical plane containing the drag force. It may increase the time of flight, and is written as

$$\text{Lift force } (\mathbf{L}) = \frac{1}{2}\rho A v^2 C_L \hat{\mathbf{n}} \tag{6.12}$$

where C_L is the lift coefficient. For an axisymmetric body the lift coefficient is proportional to $\sin \delta$, and for small δ it may be approximated simply by δ times a constant.

Since ρ decreases with height above sea-level, equation (6.12) explains why helicopters and planes have ceiling heights. At these heights, according to equation (6.12), they can no longer develop a lift greater than their weight and so cannot continue to rise.

For many projectiles the sideways effect is negligible but when it is significant it takes a form similar to that of the lift force as

$$\text{Sideways force (S)} = \frac{1}{2}\rho A v^2 C_S(\hat{\boldsymbol{\tau}} \times \hat{\mathbf{n}})$$

where C_S is the sideways coefficient. The effect is considerable for rapidly spinning anti-aircraft shells fired almost vertically.

Example 6.2

A Nordic ski-jumper with his skis on has a mass of 70 kg, and his frontal area in the jumping position through the air is 0.7 m². At a certain jumping altitude the air density is 1.007 kg m^{-3} and the value of g is 9.81 ms^{-2}. If the drag and lift forces are proportional to the square of the speed with $C_D = 1.78$, $C_L = 0.44$ and if his velocity at the take-off point is 15 ms^{-1} in the horizontal direction, determine where he will land on the hill-slope whose angle is 45°. The take-off point is 10 metres vertically above the slope.

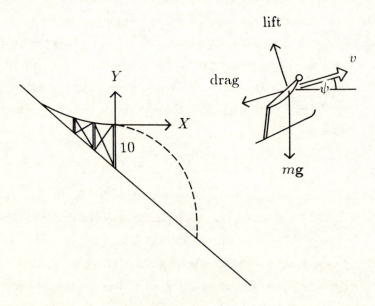

Figure 6.5 Force diagram and jump profile for Nordic ski-jump

Solution

With the take-off point as origin, the equation for the hill-slope is

$$y = -x - 10 \tag{6.13}$$

If the only major forces acting are assumed to be gravity, drag and lift, the governing equations in the tangential and normal directions may be written, with the help of equation (3.4), as

$$\left.\begin{aligned}
m\frac{dv}{dt} &= -mg\sin\psi - \frac{1}{2}\rho v^2 A C_D \\
mv\frac{d\psi}{dt} &= -mg\cos\psi + \frac{1}{2}\rho v^2 A C_L
\end{aligned}\right\} \tag{6.14}$$

$$\left.\begin{aligned}
\frac{dx}{dt} &= v\cos\psi \\
\frac{dy}{dt} &= v\sin\psi
\end{aligned}\right\} \tag{6.15}$$

with $v = 15$, $\psi = 0$, $x = 0$, $y = 0$ when $t = 0$. Using the numerical values given in the problem, equations (6.14) can be rewritten as

$$\frac{dv}{dt} = -9.81\sin\psi - 0.00899v^2$$

and

$$\frac{d\psi}{dt} = \frac{-9.81\cos\psi + 0.00223v^2}{v}$$

(Method 1)

These can be solved numerically using a Runge-Kutta technique with time-step 0.01 seconds. The values at each stage are substituted into equations (6.15) which are also solved numerically to give (x, y) values at each t value. The programme is made to stop when the magnitude of the calculated y exceeds the calculated x by 10, for then the skier will have reached the slope according to equation (6.13). The appropriate landing values are calculated to be $x_L = 46.6$ and $y_L = -56.6$, and so the length of this jump when measured down the slope from the bottom of the take-off tower is

$$\sqrt{(x_L)^2 + (y_L + 10)^2} = 65.9 \text{ metres}$$

(Method 2)

Equations (6.14) and (6.15) are non-dimensionalised using the usual transformation (1.12). This produces

$$\frac{dV}{dT} = -\sin\psi - 0.206V^2$$

$$V\frac{d\psi}{dT} = -\cos\psi + 0.051V^2$$

$$\frac{dX}{dT} = V\cos\psi$$

$$\frac{dY}{dT} = V\sin\psi$$

with $V = 1$, $\psi = 0$, $X = 0$, $Y = 0$ when $T = 0$. These can be solved using the perturbation procedures of Chapter 5 with $\epsilon = 0.206$. The term incorporating the lift effect is treated in the same way as for drag. This is left as an exercise for the reader.

6.6 Wind Corrections

The wind is a turbulent movement of air which often blows in gusts and is certainly never uniform in magnitude or direction. The presence of obstacles such as man-made structures, forests and non-uniformities of the Earth's surface produces values of wind speed and direction that vary markedly not only in time but also in space. Nevertheless an average direction and strength is sometimes assigned to the wind measured at a particular level, as for example in meteorological reports. One of the most striking features of this average is that it decreases significantly as the measuring level gets closer to the ground. This is known as the boundary layer effect, which for thermally neutral conditions predicts a logarithmic velocity profile for the equilibrium wind. To make wind effects even more complicated, there is an Ekman effect due to the Earth's rotation which tends to swing the direction of the wind around in a spiral as greater heights are reached.

Calculations that include the wind can therefore become quite complicated. When a wind $\mathbf{w} = [w_x, w_y, w_z]$ is blowing, a projectile's air speed is given by

$$\mathbf{v}^* = \mathbf{v} - \mathbf{w}$$

and the drag, lift and sideways effects will all depend directly on $\mathbf{v}^*$ instead of on $\mathbf{v}$. That is, these effects are only functions of $\mathbf{v}$ when the wind is zero. The wind usually causes the angle of yaw to change, resulting in continual alterations to C_D, C_L and C_S (Section 6.5). For a cross wind the values of C_S can be quite large, and have a very significant effect, as for example with fin-stabilized projectiles where the sideways effect of wind on these fins can produce large deviations in the flight path from the no-wind case. For a headwind or tailwind, the other coefficients C_D and C_L may vary significantly with either advantageous or detrimental results. For example, athletes using the discus or javelin prefer to throw into the wind to get extra distance from lift. On the other hand, jumps which exceed the long jump record are not recognised when the tailwind is greater than 2 ms^{-1}.

The addition of the effect of the wind to the basic equation (4.1) yields

$$m\frac{d\mathbf{v}}{dt} = m\mathbf{g} - \frac{1}{2}\rho A C_D |\mathbf{v}^*|^2 \hat{\mathbf{v}}^* \qquad (6.16)$$

with $\mathbf{v} = \mathbf{v}_0$, $\mathbf{r} = 0$ when $t = 0$.

The origin is now shifted to co-ordinates fixed in the moving air mass using the transformation

$$\mathbf{r} = \mathbf{r}^* + \int_0^t \mathbf{w}\, dt \qquad (6.17)$$

This transformation is based on the relative velocities relationship $\mathbf{v} = \mathbf{v}^* + \mathbf{w}$ where $\mathbf{v}^* = d\mathbf{r}^*/dt$, and changes equation (6.16) into

$$m\frac{d\mathbf{v}^*}{dt} = m\mathbf{g} - \frac{1}{2}\rho A C_D |\mathbf{v}^*|^2 \hat{\mathbf{v}}^* - m\frac{d\mathbf{w}}{dt} \qquad (6.18)$$

with $\mathbf{v}^* = \mathbf{v}_0 - \mathbf{w}_0$, $\mathbf{r}^* = 0$ when $t = 0$. Equation (6.18) can be solved for $\mathbf{r}^*$, using either the numerical or perturbation techniques contained in the previous two chapters, which treated problems without wind. Now, however, a modified initial velocity $\mathbf{v}_0 - \mathbf{w}_0$ must be used. To finally obtain $\mathbf{r}$ from $\mathbf{r}^*$ equation (6.17) is used. This technique will be applied in Example 6.3 below.

For gunnery purposes use is made of meteorological information which gives the horizontal wind velocity at different layers above the ground.

Suppose that the y-depth is divided into N layers which have horizontal wind velocities in the x-direction of firing given by $w_1, w_2, \ldots, w_N$ respectively.

The range effect of a unit wind in the x direction is calculated numerically for each given initial velocity w_i, and is denoted by $(\Delta x)_i$. These will usually be available in tables for the gunnery officer.

Therefore the total range effect due to the N wind layers is

$$\sum_{i=1}^{N} w_i (\Delta x)_i$$

An average wind (or ballistic wind $w^{(b)}$) can be computed from this by dividing by $(\Delta x)_1$ the range effect in the bottom layer. Thus

$$w^{(b)} = \sum_{i=1}^{N} w_i \frac{(\Delta x)_i}{(\Delta x)_1}$$

$$= \sum_{i=1}^{N} w_i q_i$$

where q_i is called the ith range wind weighting factor. Using these q_i's the ballistic meteorological station can compute the ballistic range wind for various representative trajectories.

Example 6.3

What is the range and time of flight to the first bounce for a soccer ball kicked off the ground at an angle of $50°$ to the horizontal with an initial speed of 28 ms^{-1} under each of the following conditions:

(i) into a headwind of 5 ms^{-1},

(ii) assisted by a tailwind of 5 ms^{-1},

(iii) when there is a cross wind of 5 ms^{-1} blowing at right angles to the initial direction of the kick?

Solution

From the details of Example 4.1 the ballistic coefficient is $c = 43$.

(i) Headwind: This can be written as $\mathbf{w} = [-5,\ 0,\ 0]$, and then $d\mathbf{w}/dt = \mathbf{0}$.

Therefore equation (6.18) minus the last term must be solved with initial conditions $\mathbf{v}^* = [28\cos 50° + 5,\ 28\sin 50°,\ 0]$ and $\mathbf{r}^* = [0,\ 0,\ 0]$.

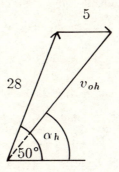

Figure 6.6 Relative velocities for a soccer kick into a headwind

In the moving frame the pseudo-speed at take-off (v_{0h}) and pseudo-angle at take-off (α_h) are therefore given by

$$v_{0h} \cos \alpha_h = 28 \cos 50° + 5$$

$$v_{0h} \sin \alpha_h = 28 \sin 50°$$

and so $v_{0h} = 31.45$, $\alpha_h = 43°$.

Hence $b_1 = v_{0h}^2 \sin 2\alpha_h / c = 22.9$. From the ballistic Table 4.2 an interpolation produces $b_2 = 1.33$ and $b_3 = 2.01$.

Therefore the pseudo-range is $(43 \times 1.33) = 57.19$ and the time of flight is $b_3 c/(v_{0h} \cos \alpha_h) = 3.76$ seconds.

Now equation (6.17) yields

$$\mathbf{r} = \mathbf{r}^* + [-5,\ 0,\ 0]t_f$$

$$= [57.19,\ 0,\ 0] + 3.76[-5,\ 0,\ 0]$$

$$= [38.39,\ 0,\ 0]$$

Therefore correct to one decimal place the range into the headwind is 38.4 metres and the time of flight is 3.8 seconds.

(ii) Tailwind: This can be written as $\mathbf{w} = [5,\ 0,\ 0]$ and again $d\mathbf{w}/dt = \mathbf{0}$.
    Therefore the ballistic Table 4.2 still applies. In the moving frame the pseudo-speed at take-off (v_{0t}) and pseudo-angle at take-off (α_t) are given by

$$v_{0t} \cos \alpha_t = 28 \cos 50° - 5$$

$$v_{0t} \sin \alpha_t = 28 \sin 50°$$

These yield $v_{0t} = 25.8$, $\alpha_t = 58.78°$, and so $b_1 = 13.7$. From Table 4.2 an interpolation produces $b_2 = 0.93$, $b_3 = 1.27$. Therefore the pseudo-range is $(43 \times 0.93) = 39.99$ and since $v_{0t} \cos \alpha_t = 13$ the time of flight is $\left(\frac{43 \times 1.27}{13}\right) = 4.20$ seconds.
    Equation (6.17) yields

$$\mathbf{r} = \mathbf{r}^* + [5,\ 0,\ 0]t$$

$$= [39.99,\ 0,\ 0] + 4.20[5,\ 0,\ 0]$$

$$= [60.99,\ 0,\ 0]$$

Hence to one decimal place the range with a tailwind is 61.0 metres and the time of flight is 4.2 seconds.

(iii) Crosswind: This can be written as $\mathbf{w} = [0,\ 0,\ 5]$ and so $d\mathbf{w}/dt = \mathbf{0}$ and the ballistic table in Chapter 4 can again be used.
    Now the velocity of the soccer ball relative to the moving frame is $[28 \cos 50°,\ 28 \sin 50°,\ -5]$. The ballistic tables are based on a trajectory in a vertical plane, and the vertical component of the initial velocity is $28 \sin 50° \hat{\mathbf{j}} = 21.45 \hat{\mathbf{j}}$, while the horizontal component is $28 \cos 50° \hat{\mathbf{i}} - 5 \hat{\mathbf{k}}$.

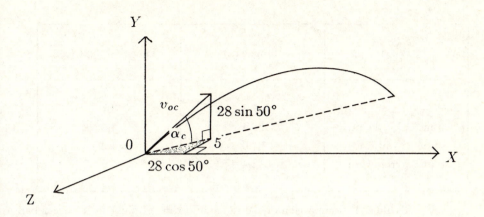

Figure 6.7 Initial velocity components in the moving frame

Therefore the pseudo-speed and pseudo-angle at take off are $v_{0c} = 28.44$ and $\alpha_c = 48.95°$, and so $b_1 = 18.6$. From Table 4.2 interpolation gives $b_2 = 1.16$ and $b_3 = 1.67$, so the pseudo-range is $(43 \times 1.16) = 49.88$. Since $v_{0c} \cos \alpha_c = 18.68$ the time of flight is $\left(\frac{43 \times 1.67}{18.68}\right) = 3.84$ seconds.

Now the pseudo-range will be measured in the direction of the initial horizontal component $28 \cos 50° \hat{\imath} - 5\hat{k}$ and so

$$\mathbf{r}^* = \left[\frac{49.88 \times 28 \cos 50°}{18.68}, \ 0, \ \frac{49.88 \times (-5)}{18.68}\right]$$

$$= [48.06, \ 0, \ -13.35]$$

Hence equation (6.17) yields

$$\mathbf{r} = \mathbf{r}^* + [0, \ 0, \ 5]t_f$$

$$= [48.06, \ 0, \ -13.35 + (5 \times 3.84)]$$

$$= [48.06, \ 0, \ 5.85]$$

Therefore to one decimal place the range is 48.4 metres, the deflection is 5.9 metres in the wind direction, and the time of flight is 3.9 seconds.

A comparison of the range and time of flight under various conditions is given in Table 6.1.

Wind	Range (m)	Time of Flight (s)
Zero (see Example 4.1)	47.3	3.7
5 ms^{-1} headwind	38.4	3.8
5 ms^{-1} tailwind	61.0	4.2
5 ms^{-1} crosswind	48.4	3.9

Table 6.1 Comparison of the range and time of flight for a soccer ball under different wind conditions

6.7 Overturning Moment

The sideways force **S** and lift force **L** for a projectile travelling with yaw combine vectorially to produce a so-called "cross-wind" force. This in turn can be combined vectorially with the drag **D** to produce the resultant aerodynamic force on the projectile. The line of action of this resultant force depends on the shape of the projectile, and the point where it cuts the axis of symmetry of a symmetrical projectile is called the centre of pressure.

Shells fired from large guns have a heavy base to protect the shell from the high pressures in the muzzle. They also have a streamlined nose to reduce forebody drag. Consequently the centre of mass (or gravity) for most shells is nearer the base than the centre of pressure. There is, therefore, a resultant moment about the centre of mass which tends to increase the angle of yaw with a related increase in the drag force. This moment is termed the overturning moment **M** and can be written for a shell as

$$\mathbf{M} = \frac{\pi}{8}\rho d^3 v^2 K_M \sin \delta \cos \delta \{\hat{\boldsymbol{\tau}} \times (\widehat{\mathbf{L} + \mathbf{S}})\}$$

where $\hat{\boldsymbol{\tau}}$ is a unit vector in the direction of the projectile's velocity, $(\widehat{\mathbf{L} + \mathbf{S}})$ is a unit vector in the direction of $\mathbf{L} + \mathbf{S}$, d is the diameter of the shell, K_M is the moment coefficient and δ is the angle of yaw (see Exercise 7, Section 7.9).

Because of the unacceptable behaviour of a projectile moving with increasing angle of yaw, techniques have been devised to counteract the effect of the overturning moment. One of these is to spin the projectile about its longitudinal axis and this will be discussed in the next chapter. Another is to shift the centre of mass of the projectile closer to the leading end, as for example with stick grenades or the javelin. A third method is to attach fins to the rear of the projectile.

Fin stabilisation is used on aircraft bombs, mortar bombs, rockets, arrows and darts. The fitting of fins increases the surface area and brings the centre of pressure to the rear of the centre of mass. Then if the nose of the projectile yaws, the resultant air pressure acting at the centre of pressure swings the projectile's nose back towards the trajectory. Finned projectiles are then said to be statically stable.

Although the overturning moment is involved directly with the angular momentum of the projectile and hence its angular velocity during flight, it has an indirect effect on the linear momentum, and hence on the range and time of flight. This is because it changes the attitude of the projectile to the air through which it is moving, so there is a change in the drag force. Since this is usually the most significant of the three aerodynamic forces, there may be significant changes to the range and time of flight.

6.8 Differential Corrections

The collective inclusion of the above effects produces projectile equations that are usually too complicated to be solved by means other than numerical computations. The presence of drag effects has already steered the analysis in this direction when treating even the basic equations for a projectile. However, when all effects other than drag and gravity are small, considerable savings in computer time result from using a theory of differential corrections which encompasses a simple addition of these effects as outlined below.

The basic equation (4.1) is rewritten as

$$\left.\begin{array}{l} \dfrac{d^2 x^{(b)}}{dt^2} = -E^{(b)} \dfrac{dx^{(b)}}{dt} \\[2ex] \dfrac{d^2 y^{(b)}}{dt^2} = -E^{(b)} \dfrac{dy^{(b)}}{dt} - g^{(b)} \end{array}\right\} \tag{6.19}$$

where a superscript (b) indicates either a parameter or variable in the basic equations and $E^{(b)} = \rho^{(b)} A v^{(b)} C_D^{(b)} / (2m)$.

Now for a projectile that is subject to Coriolis, gravity, wind, aerodynamic and atmospheric variations the above equations should be replaced by

$$\left. \begin{aligned} \frac{d^2 x}{dt^2} &= -E \frac{dx}{dt} + F_x \\[2mm] \frac{d^2 y}{dt^2} &= -E \frac{dy}{dt} - g + F_y \\[2mm] \frac{d^2 z}{dt^2} &= -E \frac{dz}{dt} + F_z \end{aligned} \right\} \qquad (6.20)$$

where $[F_x, F_y, F_z]$ is the resultant of all the proper and pseudo-forces on the projectile other than gravity. If the variations are small it is best to write

$$E = E^{(b)} + E^{(s)}$$

$$g = g^{(b)} + g^{(s)}$$

$$x = x^{(b)} + x^{(s)}$$

$$y = y^{(b)} + y^{(s)}$$

$$z = z^{(s)}$$

where the superscript (s) denotes a small quantity correction. Substitution of these into equations (6.20) and subtraction of equations (6.19) produces

$$\left. \begin{aligned} \frac{d^2 x^{(s)}}{dt^2} + E^{(b)} \frac{dx^{(s)}}{dt} &= -E^{(s)} \frac{dx^{(b)}}{dt} + F_x \\[2mm] \frac{d^2 y^{(s)}}{dt^2} + E^{(b)} \frac{dy^{(s)}}{dt} &= -E^{(s)} \frac{dy^{(b)}}{dt} - g^{(s)} + F_y \\[2mm] \frac{d^2 z^{(s)}}{dt^2} + E^{(b)} \frac{dz^{(s)}}{dt} &= F_z \end{aligned} \right\} \qquad (6.21)$$

where any product of two small quantities has been neglected. The only unknown quantities in equations (6.21) are derivatives of $x^{(s)}$, $y^{(s)}$ and $z^{(s)}$.

Now $x^{(b)}$ and $y^{(b)}$ satisfy the initial conditions $\mathbf{r} = \mathbf{0}$ and $d\mathbf{r}/dt = [v_0 \cos \alpha, \ v_0 \sin \alpha, \ 0]$, consequently $\mathbf{r}^{(s)} = d\mathbf{r}^{(s)}/dt = \mathbf{0}$ when $t = 0$. Therefore the corrections to be made to the zeroth-order approximation $[x^{(b)}, \ y^{(b)}, \ 0]$ of the projectile's position are given by the numerical solution of equations (6.21) subject to homogeneous initial conditions. A large number of results can be calculated for different values of v_0, α, the latitude λ, the altitude h and the wind $\mathbf{w}$ from which tables of corrections may be produced. These have been tabulated in great detail for gunnery.

When forces other than gravity and drag are so large that they cannot be treated as differential corrections (for example when rocket thrust is present) the equations (6.20) of motion have to be solved numerically as they stand. An example of a numerical scheme which is very efficient for these types of projectile problems is given by Goodyear (1973).

6.9 Exercises

1. A shot is fired from a point on the Earth's surface so that the horizontal component of the initial velocity points due North. If drag and variations in gravity may be neglected, show that when the shot again reaches the Earth's surface its deviation from the Northerly direction due to the Earth's rotation is to the East or the West according as $(3 \tan \lambda - \tan \alpha)$ is positive or negative, where λ is the latitude and α is the angle of projection above the horizontal.

2. At latitude 60° North a projectile is fired at an angle of 45° to the horizontal with muzzle speed 500 ms^{-1}. Consider the two cases in which it is initially fired

    (i) along a meridian of longitude in the direction North,

    (ii) along a circle of latitude in the direction West.

    Assuming that the only real force acting on the projectile is gravity, calculate for both cases the time to reach the greatest height and the deflection at that time due to Coriolis effects.

3. Find the approximate z-component of gravitational attraction of the whole Earth on a unit mass at latitude λ on the Earth's surface, when the Earth is considered as a non-rotating oblate spheroid.

4. For a projectile influenced mainly by gravity and relatively small (but not

insignificant) drag and lift forces, derive expressions for the dimensionless horizontal and vertical co-ordinates as functions of the dimensionless time, given that the initial angle of projection is zero.

5. Solve Example 6.2 using a perturbation analysis.

6. For a projectile influenced mainly by gravity, but with a relatively small drag effect and a similarly small horizontal headwind acting on it, derive an expression for the range on the horizontal plane through the projection point.

7. Find out the range of variations for temperature and pressure in your town or city. A soccer ball of mass 0.43 kg and diameter 0.22 m, for which the drag coefficient is 0.45, is kicked with initial velocity 35 ms^{-1} at an angle 45° to the horizontal. Determine the horizontal range of this kick for two different days; one when the temperature is highest and pressure lowest, and another when the temperature is lowest and pressure highest.

8. The soccer ball in the previous exercise is kicked on a day when the air is dry and its pressure is 1000 millibars. If it was similarly kicked on a day when the saturated vapour pressure of water has risen to 23 millibars determine the alteration in the horizontal range of the ball.

9. Consider a long jump with take-off speed 10 ms^{-1} and a take-off angle 20°. A cross-wind of 2 ms^{-1} is blowing at right angles to the jump direction. Show that the drag has only a small effect on the range and that there is the same alteration due to wind correction with or without drag.

10. Calculate the height of the jump of a kangaroo with $\epsilon = 0.2$ and take-off velocity given by $v_0 = 30$ ms^{-1}, $\alpha = 25°$ when it jumps into a 5 ms^{-1} headwind.

11. For the long jump initial conditions in Exercise 9 and a sinusoidal tail wind given by $\mathbf{w} = (2\sin 2\pi t, 0, 0)$ during the jump, calculate the length of the jump on the horizontal plane through the projection point.

7. SPIN EFFECTS

"_____, when I remembered that I had often seen
a tennis ball, struck with an oblique racket, describe
such a curveline. For, a circular as well as a progressive
motion being communicated to it by that stroke, its parts
on that side, where the motions conspire, must press and
beat the contiguous air more violently than on the other,
and there excite a reluctancy and reaction of the air
proportionably greater ....."

Sir Isaac Newton (1642 - 1727)

7.1 Overcoming Yaw

When a non-spherical projectile is fired or launched its longitudinal axis
rarely lies along the tangent to its trajectory, so there is a non-zero angle of yaw. As
a result of yaw the air acts on the projectile to produce

    (i) an increased drag compared with the drag at zero yaw,

   (ii) a deviating force which tends to alter the projectile's trajectory,

  (iii) an overturning moment which rotates the projectile about its centre of
       gravity.

The last was discussed in Section 6.7, and may increase or decrease the
yaw depending on the design of the projectile. A decrease in yaw is obtained by
fitting fins to some projectiles (darts, arrows, bombs and mortar shells), so providing
a method of stabilising the projectile and its trajectory. This method of stabilisation
has some disadvantages however, the principal one being that a cross-wind tends to
interact with the fins and push the projectile well off course.

The streamline design of a shell to reduce forebody drag and the need to
have a thick base to withstand the high pressure in the gun has resulted in shells with
a centre of gravity near the base and a centre of pressure near the nose. As this is the
opposite of the positions for a fin-stabilised projectile, other ways of achieving stability
are sought. The most useful method is to spin a projectile about its longitudinal axis

by an amount which is just sufficient to maintain a small angle of yaw. This spin
is imparted to shells by the rifling of the barrel, but for other projectiles it may be
imparted by other launching techniques (for example, the "torpedo" punt in football
or the javelin throw with spin produced by the releasing hand).

7.2 Spin Stabilisation of Shells

The principles involved in spin stabilisation are those of the gyroscope
or spinning top which display the important characteristics of spatial rigidity and
precession. Spatial rigidity is that property of a gyroscope which maintains its axis of
spin near the original direction of this axis. Precession may be thought of as a circular
yaw about the centre of gravity. If a spinning top is acted on by some force tending to
change its axis of spin then it will precess in a direction normal to the plane containing
that force and the spin axis, (Figure 7.1). This also occurs for spinning shells where
the force tending to change the axis of spin is a combination of gravity and drag.

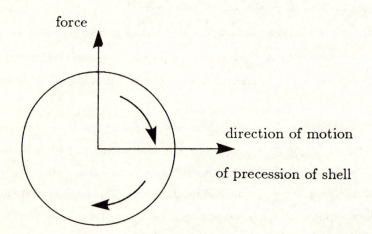

Figure 7.1 Precession direction

As a spinning shell precesses its leading tip (or nose) develops a small
circular motion superimposed on the precessional circular motion. The nose thus
moves in a rosette pattern known as a nutation. Because both nutation and precession

are rapidly damped out by spin, a spinning shell can be stabilised in flight after a very
short distance.

Consider a shell with right-hand spin (that is, clockwise motion about
the axis of symmetry looking from base to tip) leaving the muzzle of a gun. Gravity
causes the centre of gravity of the shell to follow a curved path but, owing to spatial
rigidity, the shell tends to retain the attitude it possessed at the instant of muzzle
ejection. Therefore, it develops an angle of yaw between its axis and the trajectory.
The resulting gyroscopic effect is that the nose will move to the right. Moreover, air
resistance now causes a sideways force tending to turn the nose further to the right.
The precession effect of the spinning shell causes the nose to drop. Again air resistance
attempts to increase the nose-down attitude but the precession effect moves the nose
to the left. This action continues throughout the trajectory, and the nose of the shell
describes a corkscrew type of curve about the trajectory.

While all this is happening, the trajectory is dipping. Spatial rigidity
tends to keep the axis of the corkscrew above and to the right of the trajectory. The
air stream coming towards the shell will therefore, on average, push the shell to the
right. Thus a right hand spun projectile will deviate to the right during its flight path,
this lateral deviation being called drift.

The ratio between muzzle speed and spin for most trajectories is fairly
critical. If the shell is spun too fast the overturning moment produced on the yawed
projectile by air resistance is insufficient to overcome spatial rigidity, and the shell will
land on its base. This is the case for a small arms bullet at extreme range and the
shell is said to be over-stabilised.

If the spin is insufficient the projectile cannot recover from more than a
relatively small yaw (2 or 3 degrees). The shell will precess too rapidly, the nose will
dip too quickly and the shell loses considerable range due to excessive yaw. In extreme
cases the shell may even start tumbling, which produces a loud whirring noise.

The stability factor s for a spinning shell is defined by (see Farrar and
Leeming, 1983)

$$s = \frac{8\pi (I_a)^2}{\rho n^2 d^5 I_t K_M \sin\delta \cos\delta}$$

where I_a is the axial moment of inertia, I_t is the moment of inertia about a transverse axis through the centre of gravity, ρ is the air density, n is the number of calibres (shell diameters) for one turn of rifling and d is the shell diameter. The angle of yaw δ and the constant K_M combine to form the overturning moment coefficient $K_M \sin \delta \cos \delta$. For stability of a shell the inequality $s > 1$ is required, and values around 1.5 are the usual criteria when manufacturing shells. Note that at ground level, where air density ρ is a maximum, s has a small value. Also v is a maximum at firing, so stability can be geared to initial conditions. Overall gyroscopic stability of a shell can be attained by varying the rifling and the length-to-diameter ratio of the shell. For a given barrel and rifling twist, there is a length of shell above which s drops below 1 because of an increase in I_t.

Parks (1978) investigated the stability of anti-riot projectiles. He found that for very small angles of yaw the pitching moment is a restoring one, but for larger angles the moment changes sign and an unspun projectile would then tumble. Spin stabilisation was tried, but the precessional motion in flight resulted in a build up of quite large angles of incidence, with additional drag and drift. He proposed an optimum rounding of the leading edge of the blunt-nosed PVC projectile which altered the pitching moment favourably and also reduced the drag.

7.3 Spinning Spheres

Many games such as basketball, volleyball, tennis and soccer are played with a spherical ball. Other balls such as those used for cricket or golf are almost spherical. Although a sphere is perfectly symmetrical about any axis, as soon as it starts to spin there is an asymmetry set up in the flow pattern of air moving past it. Such an asymmetry is manifested in various forces which act on the sphere causing it to swerve to the left, right, up or down.

Before considering some of these forces it is useful to examine the principle of angular momentum applied to the rotational motion of the sphere. This yields a component along the spin axis given by

$$I \frac{d^2\theta}{dt^2} = L$$

where I is the moment of inertia of the spherical ball about its spin axis $\left\{ \frac{2}{5} m(a^2 - b^2) \right.$

for a hollow ball of mass m with outside radius a and inside radius b}, $d^2\theta/dt^2$ is the angular acceleration of the ball at any instant, and L denotes the sum of the moments of the forces and couples acting on the ball about the spin axis. Because of air friction this sum of moments is in general negative after the initial launching of a projectile with spin, and consequently the angular velocity of a spinning sphere is continuously diminished during the flight path.

7.4 Fluid Dynamics

The air flow past a spinning sphere can be quite complicated. In order to understand the various effects of spin, it is necessary to come to grips with some basic ideas in fluid dynamics. Only an elementary outline is provided here, and readers who seek more detailed explanations are advised to consult one of the standard fluid dynamical texts, such as Batchelor (1967). First of all flow past a non-spinning sphere is considered and the effects of spin are introduced later.

To investigate the effect of the air flowing past a moving sphere an observer is placed on the sphere and considers the flow patterns for different sphere speeds. When air is flowing slowly past a non-spinning sphere the flow pattern is approximately symmetrical as in Figure 7.2. The lines shown in the figure are called streamlines and for steady flow indicate the paths of the air particles.

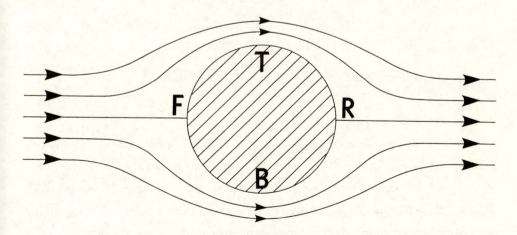

Figure 7.2 Slow-moving air past a fixed non-spinning sphere

These streamlines are more widely separated at the front and rear (F, R) of the sphere than at the top or bottom (T, B) or for that matter at the left or right sides which are not shown because Figure 7.2 is a vertical cross-section. Thus the fluid is more confined as it flows past the widest part of the sphere facing the air stream and, much like the flow in a river through a narrow section, the air speeds up in this region. As it flows towards the rear of the sphere it slows down again to its upstream value. From many experiments with fluids it is known that the air obeys Bernoulli's equation which essentially states that as the air speed increases then the air pressure drops, and vice versa. The important point to note is that as the air travels from (T, B) to R it is being slowed down near the surface of the sphere by an adverse pressure gradient.

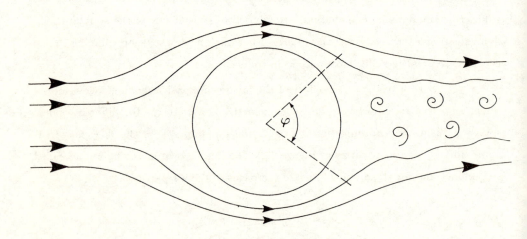

Figure 7.3 Eddying wake and separation angle φ

The behaviour is now examined when the air speed past the non-spinning sphere is increased. The first effect noted in experiments is that the flow pattern loses its fore and aft symmetry because of the action of frictional forces (viscosity). The effects of these frictional forces are at first confined to a thin boundary layer around the surface of the sphere but, as the air speed is increased still further, the retarding effect of the frictional forces and the adverse pressure gradient along the rear of the sphere together slow down the air in this region so much that, relative to the sphere, the air reaches zero speed, and then small pockets of reverse flow develop. As a result of these processes, the flow pattern behind the sphere develops a series of eddying motions in a region termed the wake (Figure 7.3). The outer surface of the wake appears to separate from the sphere's boundary layer at separation points which subtend an angle φ say at the centre of the sphere. The value of φ varies with the air speed; when the air speed is very low, φ is almost zero and the flow pattern in Figure 7.3 approaches that in Figure 7.2. But as the air speed climbs, the separation points move upstream along the surface of the sphere so that in extreme cases φ is greater than $\pi/2$ radians. Thus as the air speed increases the wake becomes wider with larger eddies, and this requires more energy. This energy must come from the kinetic energy of the ball, and hence the ball slows down because of increased drag.

But it is a peculiarity of flow around spheres (and other convex bodies) that at some critical air speed the flow in the previously laminar boundary layer changes to turbulent motion and draws in fluid from the free stream outside. This mixing of free stream fluid with boundary layer fluid allows the air more easily to overcome the adverse pressure gradient in the boundary layer. Consequently when the critical speed is reached the separation points suddenly move further back along the sphere, the width of the wake is reduced and the drag force is decreased. The critical air speed for the onset of this turbulent boundary layer depends inversely on the diameter of the sphere. Fluid dynamicists take account of this by non-dimensionalising the variables so that, if U is the ball's speed and a is its radius, a special non-dimensional parameter called a Reynolds number is introduced and defined by

$$Re = \frac{Ua}{\nu}$$

where ν is the kinematic viscosity of the fluid medium through which the ball is travelling. Then the flow pattern around a sphere of radius a moving with speed U is the same as for any other sphere moving with the same Reynolds number (for example, a sphere of radius $4a$ with speed $\frac{1}{4}U$ in the same medium). In effect, when the Reynolds number reaches a certain critical value a turbulent boundary layer is produced. This is not the whole story, since experiments show that the onset of turbulence depends to a greater degree on the roughness of the surface of the sphere. That is, critical air speed is important but surface roughness is even more so. An increase in surface roughness by a small amount can reduce the critical Reynolds number by a factor of 3.

The balls used in most games have a rough surface. During the flight of a well-kicked soccer ball its speed is well above the critical value with the consequent advantage of reduced air drag. However, for table tennis and squash balls the speed is below the critical value. If a golf ball were smooth the critical value would be greater than the driving speed and the flow would be laminar. But the dimples on the surface reduce the critical value to below the driving speed, and the ball carries further.

7.5 The Magnus Effect

Consider air flowing past a sphere spinning about a fixed axis normal to the paper as shown in Figure 7.4. The flow lines near the top of the sphere are swept along by the spin and have a separation point well to the rear. This is not so at the bottom of the sphere where the spin acts opposite to the flow line motion and results in an early separation point. The wake is therefore deflected asymmetrically down, and the lift force is positive. When the wake is deflected up due to spin in the opposite direction, the lift force is down. Earlier explanations of this effect proposed that the air speed at the top of the sphere is much greater than at the bottom and consequently by Bernoulli's theorem the pressure at the top is lower than the pressure at the bottom. This pressure difference manifests itself in a force which is transverse to the original flow direction and in this case causes the spinning sphere to rise. If the spin was about some other axis, such as the vertical axis, the spinning ball would move transversely to the left or the right depending on the direction of spin. The effect is known as the Magnus effect, and occurs in many ball games.

For example in soccer it is used in a corner-kick or to "bend" a free kick around a wall of defenders and hopefully into the goalmouth. Another example is in baseball where pitchers can impart enough spin to the ball to cause it to swerve late in its flight towards the batter. A similar effect can be obtained in cricket, and swerve is usually produced by a slow bowler. However, the swing of a cricket ball by a medium fast bowler is not a Magnus effect, and will be discussed in the next section. Although the Magnus effect can be used to great advantage in such other sports as tennis and table tennis, it can also have a detrimental effect as for example in golf when players develop a "hook" or a "slice".

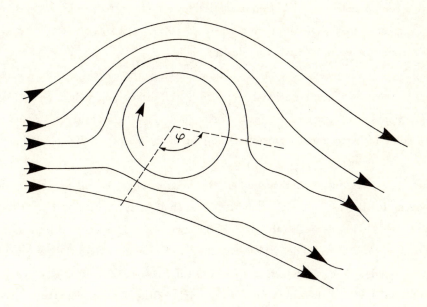

Figure 7.4 Flow lines around a spinning sphere

Note that the Magnus force is a maximum if the spin axis and the air flow direction are at right angles. The effect diminishes as these two directions approach each other, being zero when they coincide.

With the axis horizontal the Magnus effect is seen to be associated with top-spin or bottom-spin. In tennis this of course causes further problems for the receiver once the ball bounces, but we will not consider this added complication. The subject is considered in detail in Daish (1972), Chapter 15.

Golf clubs are designed so that bottom spin is given to the golf ball on all lofted shots. The Magnus force therefore becomes purely a lift force which increases the time of flight and the range. If however the ball is hit off-centre when viewed from above, the Magnus force has a component to the left or right producing a trajectory path designated as a "hook" or a "slice". Although mostly an unwanted characteristic this can be used to advantage when a cross-wind is blowing across the direction of motion of the ball. The cross-wind and the direction of motion of the ball combine to give an inclined air velocity relative to the ball. Consequently the Magnus force will have a component in the direction of motion for an off-centre "hook" or "slice" drive, and the range may therefore be increased. Even on days without wind the Magnus effect could be useful in driving a "dog-leg" hole.

The magnitude of the Magnus force in the early stages of a golf drive almost balances the gravitational force, so a well-driven golf ball appears to travel approximately in a straight line for the rising part of its flight path. This would not be so if golf balls were smooth, since the dimpling of the surface of the golf ball enhances the Magnus effect.

For non-spherical bodies, the Magnus effect is still present to varying degrees. In the case of a right-hand spinning shell from a gun, the nose of the shell points slightly above and to the right of the trajectory. There is thus a Magnus force causing movement to the left. However, this force is smaller than the opposing cross-force caused by the yaw which was discussed in Section 7.2, and in most cases for a right-hand spinning shell the drift is to the right of the plane of departure.

The most common non-spherical ball used in games is the rugby (or Australian Rules) football. When kicked in a "torpedo punt" it spins about its longitudinal axis and if this is in the direction of motion there will be no Magnus effect. However, as soon as some yaw develops there will be a Magnus force and, particularly in the case of cross-winds or head-winds, this can be effective in causing the ball to drift to the right or the left. This is also true for passes by the quarterback in American football.

The incorporation of the Magnus effect into the equations of motions for a projectile is accomplished through the values of C_L and C_S (see section 6.5), which increase appreciably as the rate of spin increases (Figure 7.5). Therefore the equation to solve is

$$m\frac{d\mathbf{v}}{dt} = m\mathbf{g} - \frac{1}{2}\rho A C_D v^2 \hat{\boldsymbol{\tau}} + \frac{1}{2}\rho A C_L v^2 \hat{\mathbf{n}} + \frac{1}{2}\rho A C_S v^2 (\hat{\boldsymbol{\tau}} \times \hat{\mathbf{n}}) \qquad (7.1)$$

where $\mathbf{g} = -g\hat{\mathbf{j}}$. For a driven golf ball, the spin may be higher than 300 rad s^{-1}, so lift or sideways coefficients of the order of 0.2 are common. Baseball pitchers and spin bowlers in cricket can only reach spins of 100 rad s^{-1} with corresponding values of C_L or C_S of about 0.05.

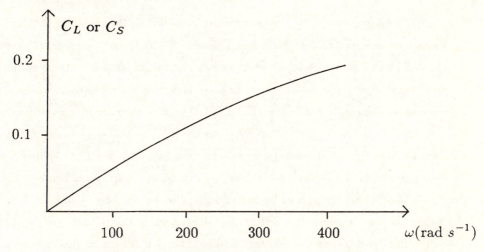

Figure 7.5 Lift or sideways coefficient versus spin rate (ω)

7.6 The Swing of a Cricket Ball

Although the effect is not caused by spin, it is appropriate at this stage to consider how the fluid dynamics discussed in Section 7.4 can explain the swing of a cricket ball. A cricket ball has the shape of a sphere with a raised equatorial boundary, called the seam, along the line where the two hemispherical pieces of leather are joined. The surface of a new ball is highly polished and at the speed of a fast bowler (above 30 ms^{-1}) the separation points are near the front of the sphere provided that the seam does not interfere. Suppose now that a seam bowler delivers the ball so that a plan view of its orientation to the relative air stream is as shown in Figure 7.6. If the hemisphere of the cricket ball to the right of the seam, as the bowler sees it, is continually polished then the separation point remains near the front of the sphere. However, the boundary layer on the hemisphere to the left of the seam can become a turbulent layer because of the seam and the lack of polishing of the ball on this side. Consequently the separation point is much closer to the rear than on the right side, and the wake moves off at an inclined angle as shown. The flow pattern in Figure 7.6 resembles that for a ball spinning counter-clockwise, and there will be a force exerted on the ball in the transverse direction from the region of high pressure R towards the region of low pressure L.

Only a small difference in pressure is needed on the two sides of the ball for there to be an appreciable swing during its flight. Under the arrangement in Figure 7.6 the ball bowled would be an out-swinger to a right-hand batsman. An in-swinger is obtained by angling the seam so that the wake is on the other side of the incoming air stream. As the ball becomes worn the polished surface becomes rougher and both sides of the ball develop a turbulent boundary layer. The tendency to swing then diminishes. Note that all the swing occurs before the ball bounces. The requirements for swing are therefore that the ball be new, that the seam be set at the correct angle during flight and that the bowler be fast enough. But for the best results he should not be too fast as he wants the ball to slow down from above the critical speed to below it as the ball travels towards the batsman. Further details on this fascinating aspect of cricket are contained in Mehta (1985) where many references are quoted.

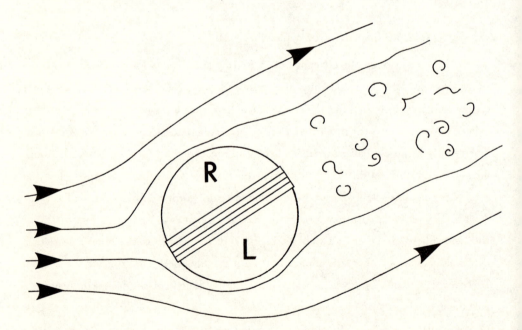

Figure 7.6 Orientation of seam for seam bowling in cricket

7.7 Spinning Seeds

Spin is used by some plants to help distribute their seeds over a large distance. Essentially the seed pod is designed like a propeller. When released from the plant a large amount of lift is generated by the propeller rotating in a horizontal plane. This lift therefore combines with the drag to overcome a large portion of the gravitational force. Consequently the time of flight is greatly increased, and if strong winds are blowing the seeds can travel quite large distances from the mother plant.

7.8 The Equations of Motion for a Spinning Shell

Using the principles of linear and angular momentum, the equations of motion for any spinning projectile in the absence of wind are

$$m\frac{d^2\mathbf{r}}{dt^2} = m\mathbf{g} + \mathbf{F}$$

and

$$\frac{d\mathbf{H}}{dt} = \mathbf{L}$$

where $\mathbf{H}$ denotes the moment of momentum about the centre of gravity. Here the only forces acting on the projectile are assumed to be gravity and the aerodynamic force $\mathbf{F}$ due to the air flow around the projectile. The moment of $\mathbf{F}$ about the centre of gravity is denoted by $\mathbf{L}$. Because of yaw the longitudinal axis of the projectile will not coincide with the direction of the tangent to the trajectory, and consequently some Magnus and precessional effects can be expected.

The general problem is extremely complex, and is one that involves 3-dimensional rotational and translational dynamics (see Synge and Griffith, 1949). The angular momentum $\mathbf{H}$ should include the moments of inertia about the principal axes of the projectile. When the axis of spin coincides with one of these principal axes, the angular momentum application is very much simplified.

To illustrate some of the complexities when spin is included, the equations of motion for the oversimplified special case of a shell spinning about its longitudinal axis and with a small amount of yaw is considered. What follows is essentially a shortened version of that given by McShane, Kelley and Reno (1953).

The linear and angular velocities of the projectile are defined respectively by

$$\mathbf{v} = v_1\hat{\mathbf{e}}_1 + v_2\hat{\mathbf{e}}_2 + v_3\hat{\mathbf{e}}_3$$

and

$$\boldsymbol{\omega} = \omega_1\hat{\mathbf{e}}_1 + \omega_2\hat{\mathbf{e}}_2 + \omega_3\hat{\mathbf{e}}_3$$

where $\hat{\mathbf{e}}_1$ is a unit vector in the direction of the projectile's longitudinal axis, and $\hat{\mathbf{e}}_1$, $\hat{\mathbf{e}}_2$ and $\hat{\mathbf{e}}_3$ form a right-handed orthogonal system.

The force $\mathbf{F}$ and torque $\mathbf{L}$ about the centre of gravity are in general functions of $\mathbf{v}$, $\boldsymbol{\omega}$, the density of air ρ, the size and shape of the projectile and the speed of sound. They may also be written in the form

$$\mathbf{F} = F_1\hat{\mathbf{e}}_1 + F_2\hat{\mathbf{e}}_2 + F_3\hat{\mathbf{e}}_3$$

and

$$\mathbf{L} = L_1\hat{\mathbf{e}}_1 + L_2\hat{\mathbf{e}}_2 + L_3\hat{\mathbf{e}}_3$$

For small angles of yaw the directions of $\mathbf{v}$ and $\boldsymbol{\omega}$ are very close to $\hat{\mathbf{e}}_1$ and the values of $v_2, v_3, \omega_2, \omega_3$ will be small in comparison with v_1, ω_1. Therefore a Taylor expansion for each component of $\mathbf{F}$ and $\mathbf{L}$ can be made yielding

$$F_j = A_j + B_j v_2 + C_j v_3 + D_j \omega_2 + E_j \omega_3$$

$$L_j = P_j + Q_j v_2 + R_j v_3 + S_j \omega_2 + T_j \omega_3$$

with $j = 1, 2, 3$, where the coefficients are independent of $v_2, v_3, \omega_2, \omega_3$. These coefficients are functions of v_1, ω_1 and are directly related to the aerodynamic coefficients C_D, C_L, C_S and the overturning moment coefficient K_M. When the projectile has a most general shape, and therefore no symmetry, it is seen that there are 30 such coefficients (five for each F_j and five for each L_j).

Suppose that the shell has the shape of a solid of revolution. From symmetry considerations, it is seen that, if the direction of yaw (δ) is reversed, the force and torque components in the longitudinal direction $\hat{\mathbf{e}}_1$ are even functions of δ, while the remaining components are odd functions of δ. Thus it is easily established by symmetry considerations that $B_1 = C_1 = D_1 = E_1 = 0$, $Q_1 = R_1 = S_1 = T_1 = 0$, $A_2 = A_3 = 0$, $P_2 = P_3 = 0$, $B_2 = C_3$, $B_3 = -C_2$, $D_2 = E_3$, $E_2 = -D_3$, $Q_2 = R_3$, $R_2 = -Q_3$, $S_2 = T_3$, $S_3 = -T_2$. This mean that there are now at most only 10 coefficients that have to be determined experimentally. Using dimensional arguments both the force and the torque due to the air flow can each be broken up into five contributions given by

$$A_1\hat{e}_1 = -\rho d^2 v_1^2 K_1 \hat{e}_1 \qquad \text{(Axial drag)},$$

$$B_2(v_2\hat{e}_2 + v_3\hat{e}_3) = \rho d^2 v_1 K_2(v_2\hat{e}_2 + v_3\hat{e}_3) \qquad \text{(Cross force due to cross velocity)},$$

$$C_2(v_3\hat{e}_2 - v_2\hat{e}_3) = -\rho d^3 \omega_1 K_3(v_3\hat{e}_2 - v_2\hat{e}_3) \qquad \text{(Magnus cross force due to cross velocity)},$$

$$D_2(\omega_2\hat{e}_2 + \omega_3\hat{e}_3) = \rho d^4 \omega_1 K_4(\omega_2\hat{e}_2 + \omega_2\hat{e}_3) \qquad \text{(Magnus cross force due to cross spin)},$$

$$E_2(\omega_3\hat{e}_2 - \omega_2\hat{e}_3) = -\rho d^3 v_1 K_5(\omega_3\hat{e}_2 - \omega_2\hat{e}_3) \qquad \text{(Cross force due to cross spin)},$$

$$P_1\hat{e}_1 = -\frac{\pi}{8}\rho d^4 v_1 \omega_1 K_6 \hat{e}_1 \qquad \text{(Spin-decelerating moment)},$$

$$Q_2(v_2\hat{e}_2 + v_3\hat{e}_3) = \rho d^4 \omega_1 K_7(v_2\hat{e}_2 + v_3\hat{e}_3) \qquad \text{(Magnus cross torque due to cross-velocity)},$$

$$R_2(v_3\hat{e}_2 - v_2\hat{e}_3) = -\rho d^3 v_1 K_8(v_3\hat{e}_2 - v_2\hat{e}_3) \qquad \text{(Cross torque due to cross-velocity)},$$

$$S_2(\omega_2\hat{e}_2 + \omega_3\hat{e}_3) = \rho d^4 v_1 K_9(\omega_2\hat{e}_2 + \omega_3\hat{e}_3) \qquad \text{(Cross torque due to cross spin)},$$

$$T_2(\omega_3\hat{e}_2 - \omega_2\hat{e}_3) = -\rho d^5 \omega_1 K_{10}(\omega_3\hat{e}_2 - \omega_2\hat{e}_3) \qquad \text{(Magnus cross torque due to cross spin)},$$

where $K_n(n = 1,\ldots,10)$ are the aerodynamic coefficients.

When $\omega_1 = 0$ all the so-called Magnus forces and torques are zero. If, in addition, $\omega_2 = \omega_3 = 0$ the only forces acting are $A_1\hat{e}_1$ and $B_2(v_2\hat{e}_2 + v_3\hat{e}_3)$ which can be resolved into the drag, lift and sideways components. The only torque acting in this special case is $R_2(v_3\hat{e}_2 - v_2\hat{e}_3)$, which is of course the overturning moment.

7.9 Exercises

1. If n is the number of calibres for one turn of rifling, v_0 is the muzzle speed of the shell, and d is its calibre, show that the initial spin of a shell as it leaves the gun is $2\pi v_0/(nd)$ rad s^{-1}.

2. Consider a cricket ball bowled at 30 ms^{-1} and suppose that it travels 15 m before bouncing. If during this part of its motion there is a difference in pressure generating a Magnus force sideways equal to $\frac{1}{4}$ of the ball's weight, determine how much lateral movement the ball undergoes.

3. A tennis ball is moving through air and its only angular velocity is a top spin. Write down the equations of motion for translational and rotational motion of the ball.

4. Assuming no wind and no yaw write down the equations of motion for a spinning seed pod.

5. For a baseball pitch thrown horizontally with a linear speed of 40 ms^{-1} and an angular spinning speed of 100 rad s^{-1} about a vertical axis use Sections 6.5 and 7.5 to determine how far the ball will slide from its initial direction of motion after it has travelled 20 metres.

6. Analyse the force and torque on a shell, which is not spinning about its longitudinal axis, which has a small amount of yaw, and which remains in its plane of departure, to show that only five coefficients are needed to describe its motion.

7. Show that

$$K_L = (K_2 - K_1)\cos^2 \delta$$

$$K_D = K_2 \sin^2 \delta \cos \delta + K_1 \cos^3 \delta$$

$$K_M = K_8$$

where δ is the angle of yaw for a spinning shell and

$$C_L = \frac{8}{\pi} K_L \sin \delta$$

8. PROJECTILES IN SPORT AND RECREATION

"The only athletic sport I ever
mastered was backgammon."

Douglas Jerrold (1803-1857)

8.1 Classes of Projectiles

So far a number of mathematical techniques have been considered that can be used to calculate projectile trajectories. With this background the particular techniques will be identified that can be applied to various projectile problems that arise in sports and recreational activities. The list of projectiles considered is not exhaustive, but is wide-ranging enough to show the general approach required if the reader wants to calculate the trajectory for a projectile that is not specifically considered in this chapter.

Two classes of projectiles will be considered; those in which the human body is the projectile and can influence the trajectory during the time of flight, and those in which the human body propels the projectile and has no further influence on the trajectory. For athletic field events the former would include the long jump, triple jump, high jump and pole vault, while the latter covers the javelin, discus, shot-put and hammer throw. The throwing events are therefore less complicated during the projectile phase, but probably more complicated during the build-up to the release of the projectile, and have been surveyed in detail recently by Hubbard (1988).

Many non-human projectiles are spherical, but other shapes will be considered also, particularly those that are streamlined.

Towards the end of the chapter three projectiles will be discussed that are somewhat unusual - boomerangs, vehicles doing jumps or plunging over cliffs, and water particles from a hose jet or water fountain. In most sections that follow the fine details of the solution will not be completed, but the main forces governing the motion will be identified and the most efficient approach will be suggested. All the references for each projectile will not be given, but at least one main reference will be provided from which other associated work can be obtained.

For sports such as tennis or soccer the ball may bounce and follow a series of trajectories before it is hit or kicked again. The analysis given here will only consider the trajectory to the first bounce and will not analyse the theory of collisions necessary to determine the release conditions for a trajectory which follows a bounce. Details of collisions in sport are contained in Daish (1972) and Stewart-Townend (1984). The assumption made here is that the general initial conditions are given for each flight path that is to be calculated.

8.2 Drag-to-weight Ratio

Most projectiles used in sporting or recreational activities are influenced by both aerodynamic forces and gravity. The techniques developed in Chapter 5 suggest that the drag-to-weight ratio should be checked for each ball game, or activity involving a projectile of some other shape, before a complicated numerical procedure is undertaken to solve the problem. In particular the value (ϵ) of the initial drag divided by the weight of the projectile should be calculated.

In Table 8.1 a list is produced of important parameters and their values for the balls and other projectiles used in a large number of sporting activities. The value assigned to the density (ρ) is taken as the value at sea-level (1.23 kg m^{-3}), but this can be modified for games played at other altitudes as indicated in Section 6.3. The kinematic viscosity (ν) of the air is 1.5×10^{-5} m^{2}s^{-1} at 20^{0}C, and remains close to this value for normal air temperatures.

The table gives typical values of the initial speed (v_0) for each projectile. A representative length scale (ℓ) is taken as the middle of the range of diameters for the spherical balls listed in the rules of each sport. For non-spherical projectiles the maximum length or width is used depending on the projectile's attitude. These values are then used to determine the Reynolds number (Re) associated with the flow of air past each projectile in flight, and hence the value of ϵ.

It appears that many ball games occur with Reynolds numbers in the range 1.0×10^5 to 2.0×10^5, the transition region where the boundary layer changes from laminar to turbulent (see Section 7.4). Although the drag coefficient (C_D) depends on both the Reynolds number and the surface roughness for non-spinning spheres,

it is assumed that the surface roughness is comparable for each ball and, as typical
values for a non-spinning sphere, $C_D = 0.45$ for ball games with $Re < 1.4 \times 10^5$ and
$C_D = 0.20$ for ball games with $Re > 1.4 \times 10^5$ are chosen.

Projectile	$A(\mathrm{m}^2)$	$m(\mathrm{kg})$	$v_0(\mathrm{ms}^{-1})$	$\ell(\mathrm{m})$	$Re = \frac{v_0\ell}{\nu}$	$\epsilon = \frac{\rho A C_D v_0^2}{2mg}$
Australian rules football	0.026	0.46	30	0.18	3.8×10^5	0.62
Baseball	0.004	0.15	40	0.07	1.9×10^5	0.53
Basketball	0.045	0.62	15	0.24	2.4×10^5	0.20
Cricket ball	0.004	0.16	35	0.07	1.5×10^5	0.38
Discus (men)	0.008	2.00	25	0.22	3.7×10^5	0.15
Discus (women)	0.007	1.00	25	0.18	3.0×10^5	0.28
Golf ball	0.001	0.05	70	0.04	1.9×10^5	1.23
Hammer	0.011	7.26	25	0.12	2.0×10^5	0.02
Javelin (men)	0.001	0.80	30	0.03	0.6×10^5	0.64
Javelin (women)	0.001	0.60	26	0.025	0.4×10^5	0.85
Long jumper	0.600	75	10	1.80	12.0×10^5	0.03
Rugby football	0.028	0.42	30	0.19	3.8×10^5	0.75
Shot-put (men)	0.011	7.26	15	0.12	1.1×10^5	0.01
Shot-put (women)	0.008	4.00	12	0.10	0.8×10^5	0.008
Shuttlecock	0.001	0.005	35	0.04	0.2×10^5	27
Soccer football	0.038	0.42	30	0.22	4.4×10^5	1.02
Squash ball	0.001	0.02	50	0.04	1.3×10^5	3.52
Table tennis ball	0.001	0.002	25	0.04	0.7×10^5	8.80
Tennis ball	0.003	0.06	40	0.06	1.6×10^5	1.00
Water polo ball	0.038	0.42	15	0.22	2.2×10^5	0.23

Table 8.1 Factors in the drag-to-weight ratio for various projectiles in
air

Note the wide range of values for ϵ. When ϵ is small enough for ϵ^2 to be
ignored (say $\epsilon < 0.2$), or large enough for $(\bar{\epsilon})^2 = \epsilon^{-2}$ to be ignored, the perturbation

techniques as outlined in Chapter 5 will be useful. This should apply to the shot-put, the hammer, the long-jumper and the shuttlecock since extremely small values for ϵ or $\bar{\epsilon}$ are obtained for all of these. It may also apply to the basketball, the water polo ball and the discus.

When the balls spin (as most of them do) the ratio of lift force to gravity force must be included, or the ratio of sideways force to gravity force. This means that there will be extra terms to consider in the equations of motion which depend on aerodynamic lift and sideways coefficients. It has been shown by Davies (1949) that the drag coefficient C_D is not altered very much by the introduction of spin to a translating sphere; however, the lift coefficient C_L changes with spin parameter $a\omega/v$ as indicated in Figure 8.1, where a is the radius of the ball and ω is its angular speed.

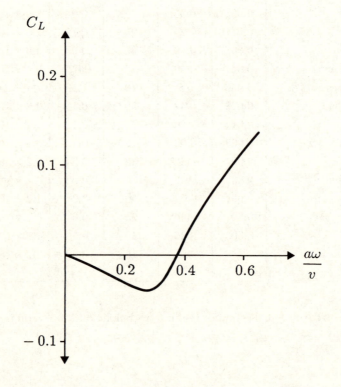

Figure 8.1 Lift coefficient for a rotating smooth sphere versus spin parameter (from Mehta, 1985)

For non-spherical projectiles also, the lift-to-weight and sideways force-to-weight ratios need to be calculated. It is only when the total aerodynamic force-to-weight ratio

$$\frac{\rho A v^2 (C_D^2 + C_L^2 + C_S^2)^{\frac{1}{2}}}{2mg}$$

is very small, or very large, that perturbation techniques are useful.

8.3 Shot-put and Hammer Throw

The shot-put develops a relative air flow with Reynolds number near 10^5, so C_D is 0.45. The value of ϵ is only 0.01, and therefore the theory of small perturbations (Chapter 5) can be used to include drag corrections. As the shot develops very little rotational motion, there will be negligible Magnus effect, and lift or sideways forces do not have to be included in the analysis.

The mathematical problem is to determine the range x in the horizontal direction for a shot projected from a height h with an initial speed v_0 at an angle α to the horizontal. The equations when non-dimensionalised become

$$\frac{dU}{dT} = -0.01 U \sqrt{U^2 + V^2}$$

$$\frac{dV}{dT} = -1 - 0.01 V \sqrt{U^2 + V^2}$$

where $U = dX/dT$, $V = dY/dT$ and the initial conditions are $U = \cos\alpha$, $V = \sin\alpha$, $X = 0$, $Y = 0$ when $T = 0$. The value of X is sought when $Y = -H = -gh/v_0^2$.

The technique used in Example 5.3 for the long jump is applicable, and the solution with $\epsilon = 0.01$ is

$$X \approx \cos\alpha \left\{ \sin\alpha + (\sin^2\alpha + 2H)^{\frac{1}{2}} \right\}$$

$$+ \epsilon \left[-\frac{1}{12}(1 + 2H)^{\frac{3}{2}} \cos\alpha + \frac{5}{8}(1 + 2H)^{\frac{1}{2}} \cos^3\alpha - \frac{2}{3} \cos\alpha \right.$$

$$- \frac{f}{2} \cos\alpha \sin\alpha + \frac{(\cos^2\alpha - 2) \sin\alpha \cos\alpha}{8f}$$

$$\left. + \left\{ \frac{\cos^5\alpha}{8f} - \frac{f}{3} \cos^3\alpha \right\} \ln \left(\frac{f + (1 + 2H)^{\frac{1}{2}}}{1 - \sin\alpha} \right) \right] \qquad (8.1)$$

where

$$f = (\sin^2 \alpha + 2H)^{\frac{1}{2}}$$

and X has an error term of order $\epsilon^2 = 0.0001$. This is a more accurate drag correction than that proposed by Lichtenberg and Wills (1978), who approximated the air resistance by linear terms. The result is dominated by the drag-free leading term, which converted back to dimensional form yields

$$x \approx \frac{v_0^2}{g}\cos\alpha[\sin\alpha + (\sin^2\alpha + 2gh/v_0^2)^{\frac{1}{2}}]$$

Therefore x increases strongly with increasing v_0, and to a lesser extent with increasing h.

One of the interesting problems in relation to the shot-put is to determine the angle of release α_m for which the component of the range in the horizontal direction (X_m) is a maximum. This occurs when $dX/d\alpha = 0$ in equation (8.1), and the predicted value to order ϵ is

$$\alpha_m = \arcsin[1/\{2(1+H)\}^{\frac{1}{2}}] + \epsilon\alpha_1(H)$$

where

$$\alpha_1(H) = \frac{(1+2H)^{\frac{3}{2}}}{2^{\frac{1}{2}}(1+H)^{\frac{1}{2}}(1+3H+H^2)}\left[\frac{(1+2H)^{\frac{3}{2}}(4H-41)}{16} + \frac{(5+2H-12H^2)}{48}\right.$$

$$-\frac{(1+5H+2H^2)}{32(1+2H)} + \frac{(3+8H+4H^2)}{64(1+2H)^3}$$

$$-\frac{(5+26H+32H^2)\{1+2^{\frac{3}{2}}H(1+H)^{\frac{1}{2}}+(1+2H)^{\frac{3}{2}}\}}{192\{2^{\frac{1}{2}}(1+H)^{\frac{1}{2}}-1\}\{1+2H+2^{\frac{1}{2}}(1+H)^{\frac{1}{2}}(1+2H)^{\frac{1}{2}}\}}$$

$$\left.+\frac{2^{\frac{1}{2}}(96H^2+50H-2)}{384(1+H)^{\frac{1}{2}}}\ln\left\{\frac{1+2H+2^{\frac{1}{2}}(1+H)^{\frac{1}{2}}(1+2H)^{\frac{1}{2}}}{2^{\frac{1}{2}}(1+H)^{\frac{1}{2}}-1}\right\}\right]$$

Substitution of the drag-corrected α_m into equation (8.1) produces the maximum range X_m. Sensitivity analyses carried out on this result show that the maximum range is altered more by changes in v_0 than by changes in α near α_m. With the above α_m the leading term for X_m is given by

$$X_m \approx (1+2H)^{\frac{1}{2}}$$

$$= H\tan 2\alpha_m$$

Wind can also be added to the analysis, as in Section 6.6, and will only affect the drag correction.

The above analysis yielding X_m may not, after all, be correct. It has been proposed by Burghes, Huntley and McDonald (1983), and independently by Hubbard (1988), that the initial speed v_0 and the angle of release α are related. In terms of non-dimensional variables, this would mean that there is a constraint relation of the form $F(\alpha, \epsilon^{\frac{1}{2}}) = 0$, since ϵ is proportional to v_0^2. The optimum value α_m for maximum X could then be determined using a Lagrange-multiplier technique. Related experiments were conducted by Red and Zogaib (1977) using a 1.14 kg metal ball, and they concluded that for a range of release angles between 26^0 and 50^0, the release speed v_0 decreases linearly with α.

A similar analysis can be used for the hammer throw, where the projectile is a shot-put ball attached to a handle by a long wire. Hubbard (1988) suggests that ϵ changes by as much as a factor of 8 from the shot to the hammer, but a more careful analysis is required because the Reynolds number for the air flow around the shot is just below the critical value, while the corresponding Reynolds number for the hammer appears to be above the critical value. Therefore, in spite of the addition of the wire and handle, the hammer could have a smaller drag coefficient than expected. It is estimated that ϵ for the hammer is only double that for the shot.

8.4 Basketball

The analysis of basketball trajectories is more important for shooting at goal than for passing. In passing, there is more emphasis on a quick direct pass without much gravity effect. However, for shooting, the effect of gravity is much more important since the aim of the game is to make the ball travel downwards through the hoop at the top of the basket. This can be done directly, or by bouncing the ball from the backboard. The former includes free shots from the foul line or higher scoring shots from any part of the basketball court, and will be mainly investigated in this section.

For a ball of diameter 24.6 cm and a hoop of diameter 45 cm there is a range of angles of entry possible for the ball to pass through the hoop. If β is the angle between the direction of motion and the horizontal as the ball passes through the hoop, elementary geometry shows that possible β values lie between 33° 8' and 90°.

Suppose x is the horizontal distance of the shooting hand from the centre of the hoop at the time of release, and that h is the height of the hoop above the throwing hand. (The height of the hoop above the floor is 3.05 metres and so h will vary with the height and action of each player). Knowing x and h, the problem in basketball is to determine the release speed v_0 and the angle of release α (see Figure 8.2).

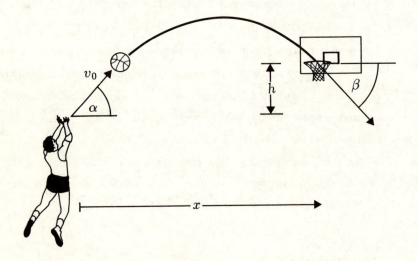

Figure 8.2 Details of a basketball shot

Since ϵ has a value 0.20 (see Table 8.1) and spin rate is low, as an initial approximation drag and lift can be ignored. From equation (1.6) with $x = v_0 t \cos \alpha$ it is seen that

$$-\tan\beta = \tan\alpha - \frac{gx}{v_0^2}\sec^2\alpha$$

since β is an angle of depression, while from equation (1.5)

$$h = x\tan\alpha - \frac{gx^2}{2v_0^2}\sec^2\alpha$$

Therefore

$$\tan\alpha = \frac{2h}{x} + \tan\beta$$

For each β lying between $33° \, 8'$ and $90°$ there are two possible values of α when h is positive. However, the larger value for α may give a trajectory that causes the basketball to hit the roof of the stadium. The speed of release is given by

$$v_0 = [gx^2\sec^2\alpha/\{2(x\tan\alpha - h)\}]^{\frac{1}{2}}$$

$$= [g(x^2\sec^2\beta + 4hx\tan\beta + 4h^2)/\{2(h + x\tan\beta)\}]^{\frac{1}{2}}$$

which for the range of possible β's predicts the variation of v_0 permissible for each shooter. Brancazio (1981) investigated margins for error in basketball shooting and concluded that high-arch shots with minimum force have the best chance of success.

8.5 Tennis, Table Tennis and Squash

For a well-hit tennis ball the Reynolds number is near 1.6×10^5 and consequently, since the surface of the ball is reasonably rough, a turbulent boundary layer is most likely produced. Therefore the drag coefficient can be assumed to be 0.20. This produces $\epsilon = 1.0$ indicating that drag is just as important as gravity in determining the flight path.

Details of the flight path for a tennis ball can therefore only be obtained numerically. Whenever topspin or backspin is put onto the ball, lift forces must also be included. The lift coefficient can be deduced from Figure 8.1 for a smooth spinning sphere, although the surface texture of the tennis ball probably makes it advisable to produce a special graph of C_L versus $a\omega/v_0$. It appears that no one has done this yet for a tennis ball.

A table tennis ball has a relatively large value for ϵ, and therefore without spin a perturbation analysis in powers of ϵ^{-1} could be attempted. However, spin plays a very big role in determining the trajectories of most shots (attacking or defensive) and, since the surface of the ball is very smooth, the lift coefficient can be obtained from Figure 8.1. A numerical solution then seems to be the only viable approach. Further complications arise because two trajectories must be calculated for each shot since the ball must bounce on the opponent's side of the table before he or she can return it. That is, volleying is not permitted as it is in tennis.

The value of ϵ for a squash ball lies between those for a tennis ball and a table tennis ball. Large deformations of the ball caused by its collision with the walls and the floor of the court persist for part of the flight path. Hence instead of using the drag coefficient for a sphere, knowledge would be required of the drag coefficient for a spinning ellipsoid of varying dimensions. Clearly the flight path would have to be determined numerically.

8.6 Badminton

The projectile used in the game of badminton is a shuttlecock. It consists of a hemisphere of cork surrounded by feathers, as shown in Figure 8.3.

Figure 8.3 The Shuttlecock

According to Peastrel, Lynch and Armenti (1980), the terminal speed of a shuttlecock is slightly less than 7 ms^{-1}. If this is used as the representative speed for non-dimensional purposes, the resulting value of ϵ is very near 1. There are four shots that are usually played in badminton, and they each have a different typical initial speed. The drop shot is hit at about 9 ms^{-1}, the underhand clear at 30 ms^{-1}, the overhead clear at 40 ms^{-1} and the smash at 60 ms^{-1}. The smash travels almost in a straight line, and with a value of ϵ in the vicinity of 9 lends itself to a perturbation analysis in powers of $\bar{\epsilon} = \epsilon^{-1}$. The other three shots can have their trajectories calculated by numerical solution of the governing differential equations. However Tan (1987) used the semi-analytic approach, as expounded in Chapter 3, to obtain trajectories that agreed well with observations.

8.7 Golf

For a golf drive there are three forces that are important, namely gravity, drag and lift. The ball is usually driven off the tee or the ground with linear speed of the order of 70 ms^{-1} and a rotational backspin of about 300 rad s^{-1}. The Reynolds number is of the order of 2×10^5, the dimples on the ball being designed so that the drag is as small as possible at these relatively high speeds. As the flight path proceeds, the ball loses both linear and angular speeds, and experiments by Bearman and Harvey (1976) have shown that the lift and drag coefficients (C_L and C_D) are not like those in Section 8.2, but change appreciably with spin parameter $a\omega/v$ (see Figures 8.4, 8.5).

Because C_L and C_D vary, the most appropriate technique for the calculation of golf ball trajectories appears to be via a numerical solution. When there is no hook or slice the equations to consider are (see equation (7.1))

$$
\left.
\begin{aligned}
\frac{d}{dt}(v \cos \psi) &= -\frac{\rho A}{2m}v^2(C_D \cos \psi + C_L \sin \psi) \\[2mm]
\frac{d}{dt}(v \sin \psi) &= -\frac{\rho A}{2m}v^2(C_D \sin \psi - C_L \cos \psi) - g \\[2mm]
\frac{dx}{dt} &= v \cos \psi \\[2mm]
\frac{dy}{dt} &= v \sin \psi
\end{aligned}
\right\} \tag{8.2}
$$

with $x = y = 0, v = v_0, \psi = \alpha$ when $t = 0$. A Runge-Kutta step-by-step approach inserting the appropriate values for C_D and C_L at each step of the solution technique can be used. If this is done for a large number of values for the initial conditions

v_0 and α, it is essentially equivalent to constructing a ballistic table. The solution produced has the typical characteristics of a golf trajectory with a slight concave upwards curvature initially.

From Figures 8.4 and 8.5 typical values for golf balls are $C_L \approx 0.25$, $C_D \approx 0.3$ and hence the value for $\epsilon_D = \rho A v_0^2 C_D/(2mg) = 1.8$ and for $\epsilon_L = \rho A v_0^2 C_L/(2mg) = 1.5$. Again this emphasises that the numerical solution appears to be the only viable technique.

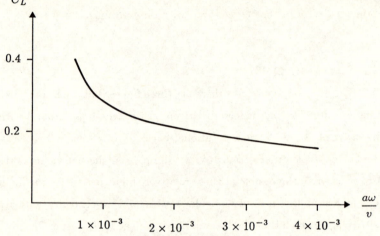

Figure 8.4 Lift coefficient versus spin parameter for a golf ball (from Bearman and Harvey, 1976)

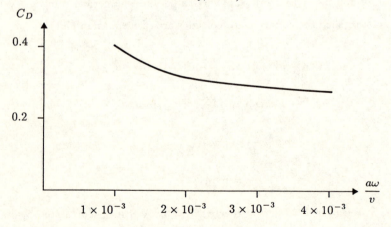

Figure 8.5 Drag coefficient versus spin parameter for a golf ball (from Bearman and Harvey, 1976)

However Tait (1890, 1891, 1893) proposed a simplified form for the drag and lift. As the flight path proceeds, the speed and spin rate of the golf ball decrease in such a way that C_L increases. Since v^2 decreases, Tait's assumption is that $C_L v^2$ behaves like $C_L^* v$, where C_L^* is a constant. For golf ball speeds greater than 40 ms^{-1} it is observed from Bearman and Harvey (1976) that although v and ω decrease the value of C_D may be taken as approximately constant. This means that the lift force has been linearised with respect to translational speed while the drag force remains non-linear, but at least it is a quadratic form only. Tait proposed further that ψ is sufficiently small over the whole trajectory of the golf ball to take $\sin \psi = \psi$ and $\cos \psi = 1$.

The equations of motion then reduce approximately to

$$\left. \begin{array}{c} \dfrac{dv}{dt} = -\dfrac{\rho A C_D}{2m} v^2 \\[2mm] v\dfrac{d\psi}{dt} = \dfrac{\rho A C_L^*}{2m} v - g \\[2mm] \dfrac{dx}{dt} = v \\[2mm] \dfrac{dy}{dt} = v\psi \end{array} \right\} \tag{8.3}$$

Writing $K_D = 2m/(\rho A C_D)$ and $K_L^* = \rho A C_L^*/(2m)$, the solution of equations (8.3) obtained for initial conditions (v_0, α) is

$$x = K_D \ln \left(1 + \frac{v_0 t}{K_D} \right)$$

$$y = \left[\alpha K_D - \frac{K_D^2}{v_0^2}(v_0 K_L^* - g) - \frac{K_D^2 g}{2v_0^2} \right] \ln \left(1 + \frac{v_0 t}{K_D} \right) + \frac{K_D(2v_0 K_L^* - g)}{2v_0} t - \frac{g}{4} t^2 .$$

Although Tait's assumptions break down badly near the end of the trajectory, the shape of the trajectory calculated in this way resembles the observed shape quite closely. Compared with the more accurate step-by-step computational process, Tait's approximate solution produces an exaggerated range and a slightly exaggerated height to the trajectory. These errors can be considerably reduced by adjusting K_D in an appropriate way. The approximate solution produces the main features of a golf ball trajectory with a roughly straight-line initial flight path and a vertex at about $\frac{2}{3}$ of the range to impact. It is a reasonable approximation to use for woods and low iron shots but, once α exceeds 20°, it becomes unreliable.

The effect of various dimple geometries on the outside surface of the golf ball has also been analysed in recent years. For a comprehensive summary the reader is referred to Mehta (1985).

As mentioned in Section 7.5, spin can have beneficial effects by increasing lift or by enabling an expert golfer to bend the ball around trees in the line of sight to the green. It can also have detrimental effects as a hook or slice, causing duffers to end up in the rough.

8.8 Cricket

During a game of cricket the flight path of the ball is of interest at two main periods of time. Firstly there is the time interval from when the ball leaves the bowler's hand and travels down the wicket towards the batsman until it is hit or missed. Some aspects of this have already been considered in Chapter 7 but will be expanded upon a little here. Secondly there is the time after the ball is hit. The usual aim of the batsman is to score as many runs as possible, and most shots are played so that the ball travels close to the ground to avoid the chance of being caught by a fieldsman. The trajectories of shots that do travel up in the air are therefore of more interest to a fieldsman, who may be trying to attempt a catch. We are not advocating that fieldsmen use mathematical calculations to determine the flight path of the ball, although this could be done if the initial behaviour off the bat is known and wind characteristics are available. In such cases, since ϵ ranges from about 0.1 to 0.4, numerical calculations would be required using knowledge of C_D and C_L. Such calculations may suggest some general trends or principles which may be useful to fieldsmen as a guide to approaching the problem of catching a ball.

The main aim of the bowler in cricket is to deliver the ball so that the batsman is deceived into missing the ball or mistiming his hit. Slow bowlers put a lot of spin on the ball as it is released. It may then curve through the air, but the major effect is to produce a deviation in the horizontal component of the ball's velocity once it bounces on the pitch. For the flight path through the air to the first bounce, the trajectory can be calculated using a numerical technique (Chapter 4), although for slower balls a perturbation analysis through Chapter 5 is possible since ϵ is sometimes near 0.1.

For a seam bowler (one who uses the seam of the ball to make it swing to the left or right) the discussion in Chapter 7 indicates that there is a small range of critical speeds over which the pressure distributions on one half of the cricket ball are due to a laminar boundary layer, while the pressure distributions on the other half of the ball are due to a turbulent boundary layer. This imbalance of pressure distributions produces a resultant side force causing the ball to deviate, or swing, to the right or the left. The critical speed at which the asymmetry appears or disappears is a function of the seam angle, surface roughness, free-stream turbulence and the spin rate of the ball (Mehta, 1985). Side forces up to about 30% of the weight of the ball are experienced. Mehta, Bentley, Proudlove and Varty (1983) computed trajectories of cricket balls using a constant value for the side force obtained from measurements due to Barton (1982) and Bentley, Varty, Proudlove and Mehta (1982).

A simple calculation based on approximate equations can be performed to see if the observed amount of deviation due to swing is possible with these measured side force values. In the horizontal direction of motion

$$m\frac{dv}{dt} = -kv^2 \tag{8.4}$$

approximately, if vertical motions are ignored. Thus

$$v\frac{dv}{dx} = -\left(\frac{k}{m}\right)v^2$$

which can be solved to yield

$$x = \frac{m}{k}\ln\frac{v_0}{v} \tag{8.5}$$

The ideal ball to bowl is one with a late swing. When this is bowled its speed is initially just above the critical range for asymmetry of boundary layers, but after it has travelled about half the length of the pitch its speed moves into the critical region. Suppose the initial speed is 31 ms^{-1} which is just above the assumed upper critical value of 30 ms^{-1}. With $k/m = 0.003$ this gives a value 10.9 metres. The pitch is 20 metres long, but since the batsman stands 1m inside one end, and the bowler delivers the ball from a point 1 m inside the other end, there is only 7.1 metres left for the ball to travel before it reaches the batsman, and it is in this part of the flight that the deviation develops.

An estimate is now needed of the time of flight. This can be obtained from equation (8.4) by solving it directly, using separation of variables, yielding

$$t = \frac{m}{k} \left\{ \frac{1}{v} - \frac{1}{v_0} \right\} \qquad (8.6)$$

When $x = 18$ the result (8.5) indicates that the speed has dropped to 29.4 ms^{-1}. Therefore from equation (8.6) the time of flight to the batsman is 0.59 seconds, and to the point where the ball begins to be affected by the side force is 0.36 seconds. Hence the side force acts for 0.23 seconds approximately, and if it is assumed that it has a constant value of 25% of the weight of the ball, a deviation of $\frac{1}{2} \times \left(\frac{1}{4} \times 9.8 \right) \times (0.23)^2 = 0.065$ metres is predicted. This is enough to deceive a batsman and perhaps make the ball touch the outside edge of the bat for a catch in the slips.

The sultriness of the day is believed to have an appreciable effect on the swing of a ball. However a considerable number of experiments surveyed by Mehta (1985) have not been able to isolate the cause. Since then there have been a number of other possible causes suggested. These include an increase in the thickness of the boundary layer, which varies as the square root of the viscosity. The roughness of the ball is then less likely to affect the laminar boundary layer on the polished side of the ball and change it to a turbulent boundary layer. Hence the swinging nature of the new ball would not seem to deteriorate as fast as it would on a dry day. There is also some suggestion that the stitches on the seam swell on a humid day, and that this would help to produce a turbulent boundary layer more easily.

8.9 Baseball

Calculations for a baseball trajectory are similar in many ways to those for a cricket ball. Measured speeds of the fastest baseball pitch and the fastest delivery in cricket show the baseball ahead. This is to be expected because of the constraints required in a legal bowling action, but the difference is not very much because a bowler in cricket is allowed to run up and bowl the ball, whereas a pitcher must keep one foot planted on the mound up to the time of delivering the pitch.

The drag coefficient is 0.20, while the lift coefficient has been determined experimentally by Watts and Ferrer (1987), and is shown in Figure 8.6.

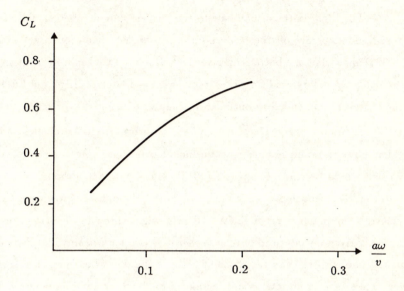

Figure 8.6 Lift coefficient versus spin parameter for a baseball (from Watts and Ferrer, 1987)

The curveball is released with top spin about the horizontal axis. The Magnus effect forces the ball to curve downwards faster than if no spin were present (Briggs, 1959). If the initial speed and spin rate are known, the trajectory can be calculated using numerical techniques.

The knuckleball is released with very little spin at all. The boundary layer is tripped by the seam of the ball so, if this is set asymmetrically initially, there will be a deviation (or swing) similar to that obtained with a cricket ball. Since the distance from the pitcher's mound to the batter is 18.3 m, the same approximate technique can be used in baseball to calculate the deviation of the ball as was used in the cricket section.

Watts and Sawyer (1975) found experimentally that large fluctuations (including a change of sign) occurred when the seam of the baseball, thrown without spin, coincided approximately with the point where the boundary-layer separation occurs. Thus there was a distinct change in direction of the ball's trajectory which could sometimes change back a short distance further on.

The flight paths of balls hit to the outfield can also be calculated if the initial speed and spin rate off the bat are known. Recent calculations by Watts and Baroni (1989) show that the batter can give the ball a large backspin by swinging just under the ball's trajectory resulting in an increased range due to increased lift. Wind variations can also be included as outlined in Chapter 6.

A similar analysis can be applied to throws from the outfield back to an infielder, but lift would not need to be included because spin rates imparted to a thrown ball would be slow. Brancazio (1984) points out that although the aim in batting a baseball is to achieve a maximum distance, the aim in throwing a baseball is to achieve a minimum time of flight. Clearly when throwing from one point to another with a given initial speed of release there are two trajectories that can achieve this, but the one with the shortest time of flight is the one whose launching angle is shallower. There is also an interesting section in his book on judging where a fielder should position himself in the field so as to catch a fly ball hit high into the air.

8.10 Soccer

Although the soccer ball is slightly smaller and lighter than a basketball, the speed with which it leaves the player's kicking boot gives it a drag-to-weight parameter ϵ very near unity. This means that trajectories for a socccer ball should be calculated using numerical integration of the governing differential equations. This is particularly so when spin is imparted to the ball to give it lift or, in many cases, to make it deviate laterally. Such deviations are used by attackers to bend the ball around a wall of defenders from a free kick and hopefully into the goalmouth, or to help a ball drop fast enough to go under the bar into the goalmouth when it is chipped over the head of the oncoming goalkeeper.

Some of the most interesting kicks from a trajectory point of view are those taken by a goalkeeper. These are usually kicked directly upfield, as deep into the opposing team's half of the field of play as possible. Not much spin is imparted to the ball in these circumstances, particularly when the goalkeeper picks the ball up and drops it onto his foot during the kicking motion. Ranges of the order of 60 to 70 metres are commonplace to the first bounce.

Wind has a large effect on the trajectory of a soccer ball, and can bring about large deviations in the horizontal direction also, particularly when spin is imparted. Another factor that affects a soccer ball's flight path is the dampness of the ball due to rain and mud, although new materials for the ground surface and the ball casing have greatly reduced this.

The value for the drag coefficient for a soccer ball is taken as 0.20 since the Reynolds number is quite large. The lift and sideways coefficients are taken as the values predicted by Figure 8.1.

8.11 Rugby and Australian Rules Football

The balls used in Rugby League, Rugby Union and Australian Rules football are approximately the shape of a prolate ellipsoid. They vary only slightly in mass, length and cross-sectional area from each other, and produce a value of ϵ typically around 0.7. The trajectory has therefore to be calculated numerically from the basic differential equations.

The longest kicks are accomplished by a "torpedo" motion when the ball rotates about its longitudinal axis which remains almost parallel to its velocity vector. The motion is therefore very similar to that of a spinning shell; it gives the ball stability in flight, and reduces any tendency to yaw. The streamline shape of these footballs ensures that the drag is reduced considerably, because the boundary layer separation points are way to the rear of the ball. The drag coefficient in this orientation will therefore be less than the value 0.2 for a soccer ball, although no measurements appear to have been published anywhere. Some lift may be generated but there is a stronger sideways force generated when a significant cross-wind blows.

For more accuracy, some Rugby penalty kicks towards the goal-posts are taken in such a way that the ball is initially placed upright and tumbles end-over-end during flight. The drag coefficient is then difficult to determine since the cross-sectional area of the ball being presented to the relatively moving air flow is oscillating between

0.028 and 0.080 m^2. Following previous discussions the cross-sectional area is kept constant at the minimum value 0.028 and the drag coefficient fluctuates by a factor of 3. Therefore, to calculate the flight path for these kicks, knowledge of the rate of tumbling is required, and experimental results are needed to yield the drag coefficient and lift coefficients at various angles of attack of the football to the airstream flowing past it.

Before leaving football a comment is needed on the forward pass in rugby. The trajectory of the ball from the passing player to the receiving player can be considered as a straight line approximately, particularly when the distance between the two players is very short. Burns (1968) noted that technically many passes in a flowing backline movement are forward.

It is possible to consider a model problem of the situation. Suppose the passer and receiver are running towards the try-line at 6 ms^{-1}, and that the receiver is 4 metres away to one side and 1 metre behind the passer (see Figure 8.7).

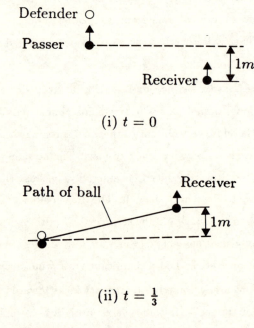

Figure 8.7 Model rugby pass with passer and receiver moving

If the speed of the pass is 10 ms^{-1} relative to the passer, then the actual speed relative to the ground is approximately 12 ms^{-1} (actually $\sqrt{136}$). The ball travels across to the receiver in approximately $\frac{1}{3}$ seconds. During this time the passer has probably slowed down to avoid running into his would-be tackler, but the receiver has travelled 2 metres forward. He is now 1 metre ahead of the initial line through the passer, and to a spectator sitting level with this line the ball has clearly travelled forward. He thinks that the referee should blow his whistle for a forward pass and halt the play. The rules clearly state that a throw-forward or forward pass occurs when the ball is deliberately propelled towards the opponents' dead-ball line. Since the component of velocity of the ball relative to the passer towards his own dead-ball line is usually less than his running speed towards the opposition's dead-ball line, the ball is propelled forwards. But the referee has been running with the play, and to him the ball has not travelled forward, and so he does not rule an infringement although technically he should.

8.12 Javelin

When analysing the motion and trajectory of a javelin (or a spear) there are extra complications not present in the motion of a spherical projectile. The elongated and asymmetrical shape ensures that the centre of pressure and the centre of mass do not coincide. Consequently, besides gravity, drag and lift, there will also be an overturning moment. This overturning moment causes the javelin to have a changing angle of attack (yaw) during the flight path, and these changes affect drag and lift considerably.

The time of flight for a well-thrown javelin can be of the order of 5 seconds. During the first 2.5 seconds the javelin has a small angle of attack with relatively low drag and lift. In the second half of its flight the angle of attack increases dramatically and finally decreases as large amounts of lift and drag are generated. The most important property of the javelin during flight is that its centre of pressure moves back and forth as the angle of attack is increased. First it is behind the centre of mass, then ahead and then behind again. Therefore the equations of motion need to include both linear and angular momentum, and the method of solution is given in great detail in the paper by Hubbard and Rust (1984), which contains a useful survey of all previous experimental and numerical research into javelin trajectories.

8.13 Discus, Frisbee and Flying Ring

The discus, frisbee and flying ring have special properties as projectiles. Their planform is a circle or the annulus of a circle, they are thin, and they are projected with a large amount of spin about the axis normal to the planform. Figure 8.8 shows their plan and mid-section profiles.

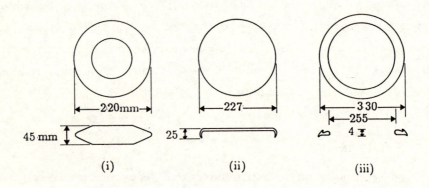

Figure 8.8 Plan and mid-section profiles for (i) men's discus (ii) frisbee
(iii) flying ring

The discus is affected by gravity, drag and lift forces. The drag and lift coefficients depend strongly on the angle of attack of the plane of the discus to the translational velocity direction. The equations to be solved numerically are (again see equations (7.1) and (8.2))

$$m\frac{d^2x}{dt^2} = -\frac{1}{2}\rho A v^2 (C_D \cos\psi + C_L \sin\psi)$$

$$m\frac{d^2y}{dt^2} = -mg + \frac{1}{2}\rho A v^2 (C_L \cos\psi - C_D \sin\psi)$$

where ψ is the angle between the direction of the translational velocity and the horizontal. The drag coefficient varies from 0.1 to 1.1 as the angle of attack changes from 0° to 90°, but at the same time the lift coefficient varies from zero to 0.9 (at 30°) and back to zero. The rotation of the discus stabilizes its orientation during flight, and in the absence of applied torques the discus' initial orientation is preserved throughout its flight path. It is important therefore that the initial angle of attack, which will be maintained, is the angle which gives maximum range.

Frohlich (1981) discussed the solution of the discus equations in detail. He indicated that if the angle of attack is initially 5° to 10° less than the initial angle of release there will be negative lift at the beginning of the flight, but the overall effect is to give minimum drag and optimum average lift throughout the upward part of the flight path. He also calculated the effect of wind and showed that throwing into a moderate wind can increase the range appreciably. He recommended that the release angle in still air should be between 30° and 40°, and this confirmed an earlier detailed calculation by Soong (1976), whose calculations also acknowledged the advantage of throwing into a head wind. But the major contribution by Soong was to show that aerodynamic torques, as well as forces, have a significant effect on the flight behaviour since they can change the attitude of the discus towards the end of the trajectory to give it more lift.

Because it is much lighter but only slightly smaller in size (see Table 8.1), a woman's discus will experience larger lift forces per unit mass than a man's discus under otherwise identical conditions. If it has the same release velocity, it can therefore be thrown further.

The same comments can be applied to the much lighter, but similar in size, frisbee. The lightness means that it can be thrown farther than a discus with the same release speed. However, unlike the discus, it appears to be in the regime of aerodynamic forces-to-weight ratio where throwing into the wind decreases the range. It is launched with a flick of the wrist to give it a large amount of spin, and gains lift through the difference of air speeds across its top and bottom surfaces (Shelton, 1975).

A new projectile has come onto the market in the last few years. This is the flying ring which is thrown flat like a frisbee and with a large amount of spin. It is a thin annulus with an aerofoil cross-section, and the distances obtained are quite impressive. Claims of at least two to three times the distances obtained by a frisbee under similar release conditions are well justified. It can almost certainly lay claim to be the projectile which can be thrown or kicked the furthest away without any other mechanical assistance; the current official record being 395.63 metres. The lift force must therefore be quite considerable and the drag force small, but no published results for C_L and C_D are available.

8.14 Long Jump, High Jump and Ski Jump

The jumps all have a phase where the human body can be treated mathematically as a projectile. During this phase the body can be moved to produce different advantageous effects. For example, a long-jumper uses a hitch-kick in flight to overcome the effects of the initial overturning moment and so land with feet ahead of the centre of gravity. On the other hand, the ski-jumper leans forward during the whole flight path until his or her head is quite near the tips of the skis.

The long jump has four phases; the run-up, the take-off, the flight as a projectile, and the landing. The main force acting during the flight phase is gravity, hence any drag or lift effects during this phase can be calculated using a perturbation procedure. Many authors have considered the governing equations containing just gravity and drag, and used either numerical procedures (Ward-Smith, 1984) or rough approximations (Brearley, 1979; Burghes et al, 1982). These were discussed in de Mestre (1986), who produced correct first-order drag corrections via perturbation theory. Assuming that the drag coefficient times the typical area $(C_D A)$ has an average value 0.36 during the flight (Brearley, 1979), the drag corrections are shown to be less than 1%. Because of the position of the body during the flight phase, the corrections due to lift are even smaller.

The most interesting investigation has been that associated with Bob Beamon's record jump at the 1968 Olympic games in Mexico City. This record still stands at the time of writing, 21 years later, and it generated a wide-ranging study of the effects of rarefied air on athletic performances. For the long jump it was shown by Brearley (1979) and Frohlich (1985) that, during the flight phase, there was only a few centimetres difference between a jump at sea-level and a jump at Mexico City. Frohlich also pointed out that the rarefied atmosphere at those Olympics enabled Beamon to run faster during the run-up phase, and launch himself with a greater speed than at sea-level. This added about 30 cm to his jump. The extra distance jumped by Beamon to set such an outstanding record (80 cm past the previous Olympic record of 8.10 metres, and 56 cm past the previous world record) was probably due to a relatively large take-off angle.

Theoretically the optimum angle for take-off for a fixed take-off speed is slightly less than 45°, due to aerodynamic effects and the centre of mass at take-off being higher than the centre of mass at landing. However most top class long-jumpers take off near 20°. The reason for this is discussed in Brancazio (1984). The horizontal velocity component at take-off could be no better than the world record sprinting rate (≈ 10 ms^{-1}), while the vertical velocity component could not exceed 5 ms^{-1}, the value needed to raise the centre of mass just over 1 metre and achieve the world high jump record. These give a launching angle near 27°. If the athlete slows down he gets a bigger launching angle, but is it worth slowing down? Clearly measurements are needed on the long jump similar to those proposed for the shot-put. The relationship between take-off speed and take-off angle is needed for each long-jumper, and with this constraint relation the optimal angle of take-off for maximum jump distance can then be determined by a Lagrange-multiplier technique. The triple jump involves a similar, but more extensive, analysis.

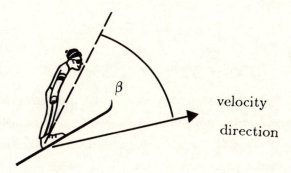

Figure 8.9 The angle of attack β for a ski-jumper

A detailed discussion of vertical jumping and the high jump, based on a point particle under the influence of gravity only, is given by Brancazio (1984). This has been extended by Hubbard and Trinkle (1985) to include an investigation of the high jump modelled as a rotating rod trying to clear a given height using various initial conditions for the rod's translational and rotational velocity. References to different types of high-jump techniques are given in both papers.

The ski jump has a ramp phase and a flight phase. Lift forces are much more important for the ski jump. In the ski-jump example in Chapter 6 it was assumed that C_D and C_L were both constant, but this approximation is too crude. Both C_D and C_L change with the angle of attack (β) between the jumper's body and his direction of velocity (see Figure 8.9).

Based on wind-tunnel tests, Krylov and Remizov (1974) suggested that

$$C_L = -0.00025\beta^2 + 0.0228\beta - 0.092$$

$$C_D = 0.0103\beta$$

The range for a ski jump can then be obtained numerically from the equations (see equations (7.1) and (8.2))

$$m\frac{dv}{dt} = -mg\sin\psi - \frac{1}{2}\rho A C_D v^2$$

$$mv\frac{d\psi}{dt} = -mg\cos\psi + \frac{1}{2}\rho A C_L v^2$$

$$\frac{dx}{dt} = v\cos\psi$$

$$\frac{dy}{dt} = v\sin\psi$$

Krylov and Remizov showed that a ski-jumper should try to ensure minimal frontal drag during the initial part of the flight, and then adjust the body angle of attack (β), so that the maximum value of lift is achieved during the second half of the flight.

Ward-Smith and Clements (1983) included an analysis of the ramp phase, then integrated the rectangular equivalent form of the above equations for a wide variety of initial conditions. They restricted the initial angle of attack to lie between 35° and 55°, as otherwise an overturning moment could develop.

8.15 Boomerangs

Excellent accounts of the physics that determines the flight of a boomerang are given by Hess (1968) and Reid (1985), while a detailed list of references is presented by Walker (1979). Essentially the boomerang behaves as a combination of aeroplane wings and a gyroscope, since it is released with a forward translational speed of the order of 27 ms^{-1} and a rotational velocity about a horizontal axis perpendicular to the initial plane of motion of approximately 60 rad s^{-1} (see Figure 8.10).

The gyroscopic behaviour produces precession and, if the leading wing (or arm) of the boomerang has a higher lift profile than the trailing wing, the boomerang will gradually change its axis of rotation from the horizontal to the vertical, and eventually lie down during flight. The lift force generated in this orientation helps to overcome the force of gravity for an appreciable time, and may result in flight times of the order of 10 seconds, with the current world record for time of flight at 151 seconds.

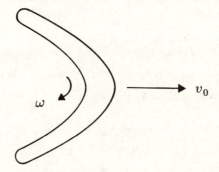

Figure 8.10 Vertical plane of release for a returning boomerang

The orbital flight path projected onto the ground approximates closely a circle. The radius of the orbit is independent of the angular spin speed, the forward speed and the length of the arms. However it increases with increasing material density of the boomerang, and decreases with increasing C_L. Under the influence of gravity the boomerang's orbit is described approximately on the surface of a large sphere. It is the diameter of the sphere rather than the diameter of the orbit that is fixed for a particular boomerang.

Although the generated lift force keeps the boomerang in the air for a longer period of time it does not contribute significantly to the flight path calculations. Hess (1975) gave a numerical solution of the equations for linear and angular momentum for a boomerang. These included gravity, average drag , the velocity of precession and the ratio of the components of the average torque on the boomerang, which gives a measure of the amount by which the boomerang turns to the left compared with the amount by which it lies down horizontally. His theoretical plots for the trajectory under different initial conditions seem to agree reasonably well with the general appearance and peculiarities of real boomerang paths.

8.16 Water Jets

Water particles from a spray, or dropped from an aeroplane to put out fires, can be considered as projectiles. Moreover when the particles are combined in a continuous flow forming a jet, their behaviour is still governed under certain conditions by projectile motion.

Discussion will be restricted here to the projectile aspects of these types of jets, and only brief mention will be made of the fluid mechanical aspects which have been investigated in detail in the references given.

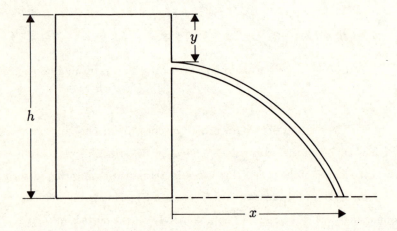

Figure 8.11 Water jet from an orifice under a constant pressure head

Consider, first of all, water flowing from an orifice under a constant pressure head (Figure 8.11). Let h be the depth of water in the vessel and y be the distance of the orifice below the water level. From Bernoulli's theorem for an ideal fluid the velocity v_0 of emergence of the jet is $\sqrt{2gy}$ in the horizontal direction. For a real fluid this is adjusted by the friction coefficient (c_v) to $c_v\sqrt{2gy}$ to account for friction losses at the orifice.

The particles in the jet are then assumed to behave as projectiles under gravity only. Hence the time of flight is given by

$$t_f = \sqrt{\frac{2(h-y)}{g}}$$

The horizontal distance x travelled by the jet when it reaches the horizontal plane through the base of the vessel is

$$x = v_0 t_f$$

$$= 2c_v\sqrt{y(h-y)}$$

Now $dx/dy = 0$ when $y = \frac{1}{2}h$ showing that the maximum distance for constant head is obtained from an orifice half-way up the side of the vessel. I am indebted to Ernst Steller, formerly of the University of Queensland, for bringing this fluid-projectile problem to my attention. In some undergraduate physics texts an incorrect illustration is produced for this problem under the heading Torricelli's theorem. The correct illustration is given in Figure 8.12.

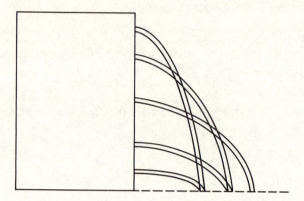

Figure 8.12 Jet pattern from a number of orifices in a constant head vessel

Flow from a hose was considered by Tuck (1975). For a sufficiently thin jet the behaviour is equivalent to a free ballistic projectile under gravity. This is the zeroth-order approximation of the solution expanded in powers of the non-dimensional thickness parameter (the inverse square of the Froude number). The first-order correction determines the shape of the cross-section of the jet at different points along its trajectory. Assuming a steady, inviscid, incompressible jet, Tuck showed that the fluid behaviour in the jet (governed by Laplace's equation and free surface boundary conditions on an unknown boundary) produces an elliptical profile with an increasing vertical major axis and a decreasing horizontal minor axis as points more distant from the source are reached. When the profile becomes too thin, instabilities due to surface tension take over, and the jet breaks up.

The flow in a waterfall was considered by Clarke (1965). The fluid mechanics solution was obtained by matching the inner (upstream of the waterfall) and outer (in the waterfall) asymptotic expansions. A more general approach for two-dimensional water jets, that included emergence of a free-flowing jet at all angles, can be found in Keller and Geer (1973).

8.17 Cars and Cycles

Vehicles can be considered as projectiles when they lose contact with the ground while travelling. Sometimes this is deliberately planned, as when a dare-devil motor-cyclist decides to jump over a number of parked vehicles for a stunt. At other times a vehicle may accidentally take off into the air, such as when a car crashes through the safety barrier on a cliff-face road. For a car travelling at 80 km h^{-1} and weighing 1000 kg the value of ϵ is approximately 0.02. Therefore its trajectory through the air is governed mainly by gravity and the initial conditions.

A real problem of this nature was discussed by Hart (1982). A car was driven off a horizontal causeway and hit a tree at a point 12 metres horizontally away from the edge of the causeway and 2.7 metres vertically below it. The driver claimed that he was doing no more than 65 km h^{-1}, and the police asked a mathematician (Professor John Blake) to comment on the claim. A calculation using equation (1.5) yields $v_0 = 58.0$ km h^{-1}, and the inclusion of the first-order correction for air drag effects still produces a value for $v_0 < 65$ km h^{-1}.

For motor-cycles, ϵ remains small, since the mass of the vehicle and rider is smaller but so is the area of cross-section. Brearley (1981) points out that the overturning moment is important during the flight if the bike-rider leaves his throttle wide open while attempting to jump over a large object. The consequences have been known to be disastrous for the stuntman, and Brearley advocates that the rider should close the throttle immediately after take-off. Such action may also be beneficial in motor-cross events, where the bikes leave the track at the top of the humps.

8.18 Seed Dispersal

In Section 7.7 mention was made of the flight of seeds using a lifting force to develop hover. Projectile motion also influences the dispersion of seeds, particularly for explosively dispersed plants such as *Hura crepitans*, a large thorny evergreen tree native to tropical America. The dispersal of the seeds from the ripe fruit of this tree has been investigated experimentally by Swaine and Beer (1977).

The equations governing the flight of any seed are those of Chapter 3 with drag proportional to v^2. From the data provided by Swaine and Beer an average ϵ for *Hura crepitans* is calculated to be approximately 3. Thus both gravity and drag are important, and the semi-analytical technique from Section 3.4 should be used. Alternatively a Runge-Kutta numerical solution can be applied, with C_D obtained experimentally for the particular seed being investigated. When a wind is blowing to help disperse the seeds, either technique can be modified along the lines suggested in Section 6.6 to predict the resulting range for a seed fired from height h with a given release speed v_0 and release angle α.

8.19 The Longest Throw or Kick

In Section 8.13 it was mentioned that the flying ring is probably the inert projectile heavier than air that can be thrown or kicked the furthest without wind assistance, without being allowed to roll down a slope after it lands, without being projected from some great height, and without any other mechanical assistance such as a sling-shot, woomera, bat or club. Therefore to conclude this chapter a list is produced of the maximum distances recorded (Russell and McWhirter, 1988) for many of the projectiles that have been considered. The list is given in Table 8.2 in decreasing order.

Projectile	Maximum distance (m)
Flying ring	395.63
Boomerang (non-returning)	200.00 (unofficial)
Frisbee	167.84
Boomerang (returning)	146.00 (flightpath > 450 metres)
Baseball throw	135.88
Cricket throw	128.60
Javelin	94.58
Football kick	91.00
Hammer throw	79.30
Discus	70.86
Shot-put	22.00

Table 8.2 Record distances for various sporting projectiles

8.20 Exercises

1. Verify Tait's approximate solution for the equations (8.3) for a golf ball trajectory.

2. A cricket ball is thrown at 45° with different speeds 15, 20 and 25 ms^{-1}. Construct a table containing range in a vacuum and range in still air. Assume a drag coefficient $C_D = 0.45$.

   Hence show that although the range in a vacuum increases as the square of the initial speed, the range in air only increases in an approximately linear manner with speed.

3. A car leaves a roadway horizontally and plunges as a projectile over a cliff to hit a tree 12 metres away horizontally and 2.7 metres vertically below the roadway. Show that the assumption of no drag for the car leads to an estimated take-off speed of 58.0 km h^{-1}.

   For a car with $\epsilon = 0.02$ and drag proportional to the square of the car speed determine the take-off speed under the same conditions when drag is included.

4. A sphere projected with initial velocity v_0 ms^{-1} at an angle α radians to the horizontal travels through a vacuum under gravity forces only, and lands on the horizontal plane through the projection point. If the speed of release is related to the angle of release by

$$v_0 = 20 - 5\alpha$$

determine the optimum conditions for maximum range.

REFERENCES

Barton, N.G. (1982). On the swing of a cricket ball in flight. Proc. Roy. Soc. London, Ser.A, **379**, 109-131.

Batchelor, G.K. (1967). An Introduction to Fluid Dynamics. Cambridge Univ. Press.

Bearman, P.W. and Harvey, J.K. (1976). Golf ball aerodynamics. Aeronaut Q. **27**, 112-122.

Bentley, K. Varty, P. Proudlove, M. and Mehta, R.D. (1982). An experimental study of cricket ball swing. Aero Tech. Note 82-106, Imperial Coll., London.

Brancazio, P.J. (1981). Physics of basketball. Am. J. Phys., **49**, 356-365.

Brancazio, P.J. (1984). Sport science. Simon and Schuster, New York.

Brearley, M.N. (1979). The long jump at Mexico City. Function, Monash Univ., Aust. **3**, 16-19.

Brearley, M.N. (1981). Motor cycle long jump. Math. Gaz. **65**, 167-171.

Briggs, L.J. (1959). Effects of spin and speed on the lateral deflection (curve) of a baseball and the Magnus effect for smooth spheres. Am. J. Phys. **27**, 589-596.

Burghes, D. Huntley, I. and McDonald, J. (1982). Applying Mathematics. Ellis Horwood, Chichester, U.K.

Burns, J.C. (1968). Private communication.

Christie, D.E. (1964). Vector Mechanics. McGraw-Hill, New York.

Clarke, N.S. (1965). On two-dimensional inviscid flow in a waterfall. J. Fluid Mech. **22**, 359-369.

Courant, R. and John, F. (1974). Introduction to Calculus and Analysis. Vol.2. Wiley, New York.

Daish, C.B. (1972). The Physics of Ball Games. English Univ. Press, London.

Davies, J.M. (1949). The aerodynamics of golf balls. J. Appl. Phys. **20**, 821-828.

de Mestre, N.J. (1977). Exterior ballistics : A useful undergraduate course in applied mathematics. Int. J. Math. Educ. Sci. Technol. **8**, 251-257.

de Mestre, N.J. (1986). The long jump record revisited. J. Aust. Math. Soc. (Ser.B). **28**, 246-259.

de Mestre, Neville and Catchpole, Ted (1988). Discrete and continuous methods in ballistics. Int. J. Math. Edn. Sci. Technol. **19**, 155-164.

Drach, J. (1920). L'équation différentielle de la balistique extérieure et son intégration par quadratures. Ann. Sci. École Norm. Sup. (3), **37**, 1-94.

Farrar, C.L. and Leeming, D.W. (1983). Military ballistics. Brassey, Oxford.

Frohlich, C. (1981). Aerodynamic effects on discus flight. Am. J. Phys. **49**, 1125-1132.

Frohlich, C. (1985). Effect of wind and altitude on record performances in foot races, pole vault and long jump. Am. J. Phys. **53**, 726-730.

Goodyear, W.H. (1973). A method of numerical integration for trajectories with variational equations. A.I.A.A. **11** (12), 1732-1736.

Hart, V.G. (1982). The car crash problem. Teaching Mathematics (Queensland Assoc. Math. Teachers). **4**, 73-75.

Hart, D. and Croft, C. (1988). Modelling with projectiles. Ellis Horwood, Chichester, U.K.

Hess, F. (1968). The aerodynamics of boomerangs. Sc. American, 219, 124-136.

Hess, F. (1975). Boomerangs : aerodynamics and motion. Ph.D. thesis, Univ. of Groningen, The Netherlands.

Hubbard, M. and Rust, H.J. (1984). Simulation of javelin flight using experimental aerodynamic data. J. Biomech. **17**, 769-776.

Hubbard, M. and Trinkle, J.C. (1985). Clearing maximum height with constrained energy. J. Appl. Mech. **51**, 179-184.

Hubbard, M. (1988) The throwing events in track and field. In "The Biomechanics of Sport II", C.L. Vaughn (ed.), CRC Press, Boca Raton, Florida.

Keller, J.B. and Geer, J. (1973). Flows of thin streams with free boundaries. J. Fluid Mech. **59**, 417-432.

Kreyszig, E. (1983). Advanced Engineering Mathematics. 5th edition. Wiley, New York.

Krylov, I.A.and Remizov, L.P. (1974). Problem of the optimum ski jump. PMM. **38**, 717-719.

Leimanis, E. (1958). Surveys in Applied Mathematics. Vol. II. Dynamics and Nonlinear Mechanics. Chap.2. Wiley, New York.

Lichtenberg, D.B. and Wills, L.G. (1978). Maximizing the range of the shot put. Am. J. Phys. **46**, 546-549.

Mayevski, N. (1982). Traité de Balistique Extérieure. Paris.

McShane, E.J., Kelley, J.L. and Reno, F.V. (1953). Exterior Ballistics. University of Denver Press, Denver, Colorado.

Mehta, R.D. Bentley, K. Proudlove, M. and Varty P. (1983). Factors affecting cricket ball swing. Nature. **303**, 787-788.

Mehta, R.D. (1985). Aerodynamics of sports balls. Ann. Rev. Fluid Mech. **17**, 151-189.

Milenski, John W. (1969). Light anti-aircraft projectile ballistics. A.I.A.A.. **7**(11), 2144-2147.

Moulton, F.R. (1962). Methods in Exterior Ballistics. Dover, New York.

Murphy, R.V. (1972). Maximum range problems in a resisting medium. Math. Gaz. **56**, 10-15.

Parker, G.W. (1977). Projectile motion with air resistance quadratic in the speed. Am. J. Phys. **45**, 606-610.

Parks, P.C. (1978), The dynamic stability in flight of spinning blunt body projectiles. Proc. AGARD Conference. Athens, Greece.

Peastrel, M. Lynch, R. and Armenti Jr., A. (1980). Terminal velocity of a shuttlecock in vertical fall. Am. J. Phys. **48**, 511-513.

Red, W.E. and Zogaib, A.J. (1977). Javelin dynamics including body interaction. J. Appl. Mech. **44**, 496-498.

Reid, R.J.O. (1985) The physics of boomerangs. Math. Spectrum. **17**, 48-57.

Russell, A. and McWhirter, N. (1988). Guiness Book of World Records. Sterling, New York.

Shelton, J. (1975). The physics of frisbee flight. In "Frisbee : A Practitioner's Manual and Definitive Treatise". Stancil E.D. Johnson (ed.), Workman Publishing Co.

Soong, T.-C. (1976). The dynamics of discus throw. J. Appl. Mech. **43**, 531-536.

Stewart-Townend, M. (1984). Mathematics in sport. Ellis Horwood, New York.

Swaine, M.D. and Beer, T. (1977). Explosive seed dispersal in *Hura Crepitans* L. (Euphorbiaceae). New Phytol. **78**, 695-708.

Synge, J.L. and Griffith, B.A. (1949). Principles of Mechanics. 2nd ed. McGraw-Hill, New York.

Tait, P.G. (1890). Some points in the physics of golf. Part I. Nature. **42**, 420-423.

Tait, P.G. (1891). Some points in the physics of golf. Part II. Nature. **44**, 497-498.

Tait, P.G. (1893). Some points in the physics of golf. Part III. Nature. **48**, 202-205.

Tan, A. (1987). Shuttlecock trajectories in badminton. Math. Spectrum. **19,** 33-36.

Tuck, E.O. (1976). The shape of free jets of water under gravity. J. Fluid Mech. **76**, 625-640.

Walker, J. (1979). Roundabout : the physics of rotation in the everyday world. W.H. Freeman and Co., New York.

Ward-Smith, A.J. and Clements, D. (1983). Numerical evaluation of the flight mechanics and trajectory of a ski-jumper. Acta Applic. Math. **1**, 301-314.

Ward-Smith, A.J. (1984). Calculation of long jump performance by numerical integration of the equations of motion. J. Biomech. Eng. **106**, 244-248.

Watts, R.G. and Sawyer, E. (1975). Aerodynamics of the knuckleball. Am. J. Phys. **43**, 960-963.

Watts, R.G. and Ferrer, R. (1987). The lateral force on a spinning sphere : Aerodynamics of a curveball. Am. J. Phys. **55**, 40-44.

Watts, R.G. and Baroni, S. (1989). Baseball-bat collisions and the resulting trajectories of spinning balls. Am. J. Phys. **57**, 40-45.

INDEX